Studies in Computational Intelligence

Volume 523

Series Editor

Janusz Kacprzyk, Warsaw, Poland

For further volumes:
http://www.springer.com/series/7092

Studies in Computational Intelligence

Volume 172

Doran Chakraborty

Sample Efficient Multiagent Learning in the Presence of Markovian Agents

 Springer

Doran Chakraborty
Department of Computer Science
The University of Texas
Austin, Texas
USA

ISSN 1860-949X   ISSN 1860-9503   (electronic)
ISBN 978-3-319-35293-0   ISBN 978-3-319-02606-0   (eBook)
DOI 10.1007/978-3-319-02606-0
Springer Cham Heidelberg New York Dordrecht London

Printed on acid-free paper

Springer is part of Springer Science+Business Media (www.springer.com)

I dedicate this work to my parents,
Late. Salil Kumar Chakraborty, and
Banani Chakraborty.

Abstract

The problem of multiagent learning (or MAL) is concerned with the study of how agents can learn and adapt in the presence of other agents that are simultaneously adapting. The problem is often studied in the stylized settings provided by repeated matrix games. The goal of this book is to develop MAL algorithms for such a setting that achieve a new set of objectives which have not been previously achieved. The book makes three main contributions.

The first main contribution proposes a novel MAL algorithm, called Convergence with Model Learning and Safety (or CMLeS), that is the first to achieve the following three objectives: (1) converges to following a Nash equilibrium joint-policy in self-play; (2) achieves close to the best response when interacting with a set of memory-bounded agents whose memory size is upper bounded by a known value; and (3) ensures an individual return that is very close to its security value when interacting with any other set of agents.

The second main contribution proposes another novel MAL algorithm that models a significantly more complex class of agent behavior called Markovian agents, that subsumes the class of memory-bounded agents. Called Joint Optimization against Markovian Agents (or JOMA), it achieves the following two objectives: (1) achieves a joint-return very close to the social welfare maximizing joint-return when interacting with Markovian agents; (2) ensures an individual return that is very close to its security value when interacting with any other set of agents.

Finally, the third main contribution shows how a key subroutine of JOMA can be extended to solve a broader class of problems pertaining to Reinforcement Learning, called "Structure Learning in factored state MDPs".

All of the algorithms presented in this book are well backed with rigorous theoretical analysis, including an analysis on sample complexity wherever applicable, as well as representative empirical tests.

University of Texas, Austin
2013

Doran Chakraborty
Supervisor: Peter Stone

Contents

Abstract VII

List of Tables XIII

List of Figures XV

1 Introduction 1

2 Background 7
  2.1 Concepts Pertaining to Reinforcement Learning 7
  2.2 Concepts Pertaining to Repeated Matrix Games 11

3 Learn or Exploit in Adversary Induced Markov
  Decision Processes 19
  3.1 LoE-AIM 20
  3.2 Results against Deterministic Agents 22
  3.3 Results against Stochastic Agents 25
  3.4 Summary 28

4 Convergence, Targeted Optimality and Safety
  in Multiagent Learning 29
  4.1 Model Learning with Safety (MLeS) 30
      4.1.1 Inputs to MLeS 31
      4.1.2 High Level Idea behind MLeS 32
      4.1.3 FIND-MODEL Algorithm 34
      4.1.4 Action Selection 36
      4.1.5 Removing Assumption 1 38
  4.2 Convergence with Model Learning and Safety (CMLeS) ... 40
  4.3 Results 44
  4.4 Summary 46

**5   Maximizing Social Welfare in the Presence
     of Markovian Agents** **49**
     5.1   Road Map to Specifying JOMA 51
     5.2   Model Learning and Exploitation of Markovian Agents
           with the Sequential Structure Assumption 52
           5.2.1   Inputs to MLEM(S) 53
           5.2.2   High Level Idea of MLEM(S) 54
           5.2.3   FIND-MODEL(S) Algorithm 56
           5.2.4   Action Selection 58
     5.3   Model Learning and Exploitation of Markovian Agents
           without the Sequential Structure Assumption 60
           5.3.1   FIND-MODEL-GENERAL Algorithm 60
           5.3.2   Action Selection 63
     5.4   Joint Optimization against Markovian Agents 64
     5.5   JOMA with Identities Unknown 68
     5.6   Empirical Validation 68
     5.7   Summary 71

6   Targeted Modeling of Markovian Agents **73**
     6.1   Algorithms and Analysis 74
           6.1.1   TOMMA(S) 74
           6.1.2   TOMMA 77
     6.2   The Surveillance Game 79
           6.2.1   Game Specifics 80
           6.2.2   Intruder Behavior 81
           6.2.3   Results for the Surveillance Game 83
     6.3   The Ticket Checking Domain 90
           6.3.1   Domain Specifics 91
           6.3.2   Results for the Ticket Checking Domain 93
     6.4   Summary 96

7   Structure Learning in Factored MDPs **99**
     7.1   Sequential Structure Learning in FMDPs 102
     7.2   General Structure Learning in FMDPs 104
     7.3   Results 106
           7.3.1   Stock Trading Domain 107
           7.3.2   System Administrator Domain 109
     7.4   Summary 110

8   Related Work **113**
     8.1   Safety, Consistency and Universal Consistency 115
     8.2   Rationality and Convergence 116
     8.3   Rationality and No Regret 117
     8.4   Guarded Optimality with Focus on Targeted Optimality... 118
     8.5   Focusing Solely on Guarantees in Self-play 120

9 Conclusion and Future Work **125**
   9.1 Summary of the Contributions 126
   9.2 Future Work 128
      9.2.1 Near Term Research Possibilities 128
      9.2.2 Long Term Research Possibilities 128

Appendix A **131**
   A.1 Computation of σ_k 131
   A.2 Proof of Lemma 4.1.1 132
   A.3 Proof of Lemma 4.1.2 134
   A.4 Achieving Safety When Assumption 1 Does Not Hold 137

Appendix B **139**
   B.1 Proof of Lemma 5.3.1 139

Appendix C **141**
   C.1 Computation of σ_k for Equation 6.2 141

References **143**

List of Tables

2.1   Payoff matrices for Prisoner's Dilemma (PD) and Chicken ..   12
2.2   Payoff matrix for Rock-Paper-Scissors   14
2.3   Payoff matrix for coordination game   16

3.1   Payoff matrices   23

5.1   Payoff matrix for Battle of Sexes (BoS)   65

6.1   Different cases of intrusion considered   82

8.1   Multiagent Learning scenarios. The row and the column
      indices represent the different agent populations modeled,
      and some of the different learning criteria that have been
      proposed in MAL to date, respectively. The entries are
      representative MAL algorithms that satisfy a particular
      criterion while interacting with a particular agent
      population. A × entry means that the particular criterion
      does not apply to the corresponding agent population. A
      blank entry means there exists no algorithm satisfying that
      particular criterion for the corresponding agent
      population.   114

List of Figures

2.1 Examples of a unichain and a multichain MDP   9

2.2 Example of the partial transition function for state (R,P) . . .   17

3.1 Results against Godfather policies in PD with $K = 3$.
Against Godfather-strict, LoE-AIM eventually learns that
it should play "cooperate" (its half of the targetable pair)
and hence converges to a payoff of 3. For Godfather-lenient,
LoE-AIM learns to optimally exploit by playing
"cooperate" frequently enough so that the history always
contains one "cooperate" action for i.   23

3.2 Results against Bully and BR in PD with $K = 3$. Both the
Bully and BR policy is to play "defect" always. Against
both of these policies, LoE-AIM eventually learns to play
"defect" and converges to a payoff of 2.   24

3.3 Results against REDVALER and WOLF-IGA in Chicken,
$K = 4$. Neither WOLF-IGA, nor REDVALER guarantees the
possible final converged Nash outcome in self-play, e.g, in
both cases, self-play generates payoffs much less than 4
showing that on numerous occasions the final converged
Nash outcome was not (4,2), the one most coveted by i.
In contrast, in all of its runs, LoE-AIM converged to the
outcome (4,2).   26

3.4 Payoff matrices of 6 games with multiple Nash equilibria.
((1,1) (0,0) (0.25,0.5)) means that the game has 3 Nash
equilibria where the probabilities of playing action A1 by
both players are respectively (1,1), (0,0) and (0.25,0.5).   27

3.5   Result against the 6 games with multiple Nash equilibria.
      Against the MAL algorithms (REDVALER and WOLF-IGA)
      and BR, LoE-AIM successfully converges to its best payoff
      of 4 on all occasions thereby demonstrating its ability to
      exploit them optimally. All the benchmarks generate lower
      average payoffs when played against these agents. Against
      the other agents, LoE-AIM does better than all the other
      benchmarks (though the average payoff is lower than 4)..... 27

4.1   Against memory-bounded agents in 3 player PD. 46

5.1   Example of how the JOMA agents agree on an ordering.
      The final ordering can be obtained by reading all the sets
      from the leaf nodes from left to right,
      i.e., $B < A < C < E < D$. 66
5.2   Comparative results in different games from gamut. The
      experiments were conducted over 3 player versions of
      each game, with two self agents and the third agent
      drawn randomly from the set of benchmark agents, along
      with minimax, self agent and arbitrary memory-bounded
      strategies of size 3. The X axis of the plot shows the
      different games while the Y axis shows the converged SW
      value for each algorithm achieved in a game as a fraction of
      the best converged SW value achieved by any algorithm for
      that game. 70

6.1   The Surveillance Game................................. 81
6.2   Cumulative reward plot for Case 1..................... 84
6.3   Cumulative reward plot for Case 2..................... 85
6.4   Cumulative reward plot for Case 3..................... 85
6.5   Cumulative reward plot for Case 4..................... 86
6.6   Cumulative reward plot for Case 5..................... 86
6.7   Cumulative reward plot for Case 6..................... 87
6.8   Model selection for TOMMA(S)-1 for Case 1. The I_0 plot
      shows the model selection for intruder I_0, and so forth...... 87
6.9   Model selection for TOMMA for Case 1. The I_0 plot shows
      the model selection for intruder I_0, and so forth. 88
6.10 Cumulative reward plot for Case 7..................... 88
6.11 Model selection for MLEM(S) for Case 1. The I_0 plot
      shows the model selection for intruder I_0, and so forth...... 89
6.12 Model selection for MLEM for Case 1. The I_0 plot shows
      the model selection for intruder I_0, and so forth. 90
6.13 The transit system with 3 stations and 8 time units........ 92
6.14 The cumulative reward plot............................ 95
6.15 Model selection for TOMMA(S)......................... 95

6.16 Model selection for TOMMA............................. 96

7.1 A DBN representation with in-degree $K = 3$. The next step values for factors X_1 and X_2 for a particular action depend only on the current step values of factor sets $\{X_1, X_2, X_3\}$ and $\{X_1, X_{n-1}, X_{n-3}\}$ respectively........................ 100
7.2 Cumulative reward plot in Stock Trading................. 108
7.3 Model selection of LSE-RMAX in Stock Trading 108
7.4 Cumulative reward plot in System Administrator 109
7.5 Model selection of LSE-RMAX(S) in System Administrator 110

Chapter 1
Introduction

A multiagent system [76] can be defined as a group of autonomous, interacting entities sharing a common environment, which they perceive with sensors and upon which they act with actuators. Multiagent systems are finding applications in a wide variety of domains including robotic teams [65], distributed control [69], data mining [73] and resource allocation [25]. They may arise as the most natural way of looking at the system, or may provide an alternative perspective on systems that are originally regarded as centralized. For instance, in robotic teams the control authority is naturally distributed among the robots [65]. In resource management, while resources can be managed by a central authority, identifying each resource with an agent may provide a helpful, distributed perspective on the system [25].

Although the agents in a multiagent system can be programmed with behaviors designed in advance, it is often necessary that they learn new behaviors online, such that the performance of the agent or of the whole multiagent system gradually improves. This is usually because the complexity of the environment makes the a priori design of a good agent behavior difficult or even impossible. Moreover, in an environment that changes over time, a hardwired behavior may often be inappropriate.

An alternative to hard wiring agents with a predefined behavior is to allow them to adapt and learn new behavior online. This brings us to the field of Reinforcement Learning (RL) [70]. An RL agent learns through interaction with its dynamic environment. At each time step, the agent perceives the complete state of the environment and takes an action, which causes it to transit to a new state. The agent receives a scalar reward signal that evaluates the quality of this transition. Well-understood algorithms with good convergence properties are available for solving the single-agent RL task (such as Q-learning [74]).

However, several new challenges arise for RL in multiagent systems. In a multiagent environment the learning agent must also adapt with the behavior of other learning (and therefore non-stationary) agents in the environment. Only then will it be able to coordinate its behavior with theirs, such that a coherent joint behavior results. This non-stationarity poses the primary challenge of learning in multiagent systems and comprises the main reason that it is best considered distinctly from single agent RL. When some or all of these entities are learning, especially about each other, we arrive at the field of Multiagent Learning (or MAL for short).

MAL is often studied in the stylized settings provided by repeated matrix games (normal form games) such as the Prisoner's Dilemma, Chicken and Rock-Paper-Scissors [56]. Repeated games of this type provide the simplest setting that encapsulates many of the key challenges posed by MAL. Specifically, they abstract away the conventional notion of *state* (situatedness) and allow one to focus purely on the impact of the agents' actions on each other's outcomes, or utilities.

Such research on MAL in repeated games typically strives to develop algorithms that can provably converge to following the optimal policy when interacting with specific classes of other agents, along with decent performance guarantees in self-play (interacting with other agents with the same behavior). For example, there is a significant volume of prior work in MAL that proposes algorithms that converge to following the optimal exploitation policy when interacting with other *stationary* agents (agents who choose their actions from a fixed distribution over their action space), while also converging to following a Nash equilibrium [55] joint-policy in self-play [13, 27].

However, requiring that the other agents in the environment all be stationary is quite restrictive. For one thing, it eliminates the possibility that any of the other agents are themselves responding to the past actions of other agents. In an attempt to address the above issue, there has been a growing body of more recent work in MAL that focuses on learning in the presence of *memory-bounded adaptive* agents, or simply *memory-bounded* agents, whose policy is a (fixed) function of some historical window of past joint-actions by all the agents [58, 59]. Though memory-bounded agents are restricted to consulting only a fixed window of past joint-actions to decide their current step action, they are still a step forward towards considering "fully adaptive agents" that use the entire history of play to decide their actions.

The goal of this book is to develop MAL algorithms that achieve a new set of goals which have not been previously achieved by any MAL algorithm. Especially we are interested in modeling agents which are more complex than memory-bounded agents. To that end we further extend focus to a particular class of agent behavior that can be modeled as *Markovian* agents. We define a Markovian agent to be one that chooses its actions as a (still fixed) function of a set of discrete feature variables computed from the joint history of play. Depending on the joint-action taken, the feature values transition in a Markovian fashion on every time step. It so happens that memory-bounded agents are indeed a special class of Markovian agents whose feature space is just the set of joint-actions from a bounded history of play.

This book takes a significant step forward in the theory of MAL as it introduces novel algorithms for modeling a comparatively more complex class of agent behavior than has been modeled to date. Furthermore for all of our main algorithms, we provide sample complexity bounds.[1]

With this motivation in mind, the two main research questions that this book answers are as follows:

- *How can a group of agents learn to maximize their payoffs over time when interacting repeatedly with another group comprised of Markovian agents, whose policies are unknown?*
- *How can they achieve the above in efficient sample complexity?*

The second goal is very important as we do not just want our algorithms to model and perform well against the set of Markovian agents, we also want them to do so with a formal guarantee on the number of samples needed.

Next, we highlight the key contributions of this book.

1. Our first contribution concerns modeling memory-bounded agents. In this regard we formally frame the problem as learning in an *Adversary Induced Markov Decision Process* (or AIM for short): a general framework for modeling Markovian agents. The concept of AIM is the foundation upon which all of our modeling techniques are built. As part of this contribution, we propose a couple of algorithms that utilize the AIM framework to model memory-bounded agents, assuming their memory size is known [19]. We also present empirical evidence of the effectiveness of these algorithms by pitting them against certain representative algorithms from the MAL and game theory literature. This contribution is presented in Chapter 3.

2. Our second contribution builds on our first contribution and proposes a novel multiagent learning algorithm called *Convergence with Model Learning and Safety* (or CMLeS [20] for short) that in a multi-player multi-action (arbitrary) repeated matrix game, is the first to achieve the following three objectives:

   - Convergence: converges to following a Nash equilibrium joint-policy in self-play (when all the other agents are also CMLeS agents);
   - Targeted Optimality against memory-bounded agents: achieves close to the best response when interacting with a set of memory-bounded agents whose memory size is upper bounded by a known value;
   - Safety: achieves an individual return very close to its security value when interacting with any other set of agents;

   This contribution is presented in Chapter 4.

3. Our third contribution focuses on modeling Markovian agents which are more general than memory-bounded agents and introduces another novel MAL algorithm, called *Joint Optimization against Markovian Agents* (or JoMA for short), that achieves the following two objectives:

   - Maximizes social welfare in the presence of Markovian agents: in the presence of Markovian agents in the population, JoMA provably achieves

---

[1] Total number of actions taken to converge to the final desired behavior.

a joint-return very close to the social welfare maximizing joint-return (for the JOMA agents). As stated earlier a Markovian agent is one that decides on its next mixed action by consulting the values of a set of discrete feature variables derived from the joint history of play. The feature vector values transition in a fashion such that their values at time $t + 1$ depend only on their values at time t and the joint-action taken at time t (Markov assumption, [60]). JOMA assumes prior knowledge of a set of possible features, called the *target set* of features, some of which are assumed to characterize the unknown policy of the Markovian agent. JOMA achieves its objective of modeling the Markovian agents with the most concise model (based on only the relevant features from the target set) via efficient exploration and in the process remains sample efficient;

- Safety: achieves an individual return very close to its security value when interacting with any other set of agents;

Along with a thorough theoretical analysis of JOMA's properties, we also present some empirical results from the gamut test-bed demonstrating its relative effectiveness compared to some of its peers from the MAL literature. This contribution is presented in Chapter 5.

4. Our fourth contribution focuses on a special case scenario of a two player repeated game against a Markovian agent. Here we propose a simplified algorithm, called *T*argeted *O*pponent *M*odeler for *M*arkovian *A*gents (or TOMMA for short), which is motivated from our findings while implementing JOMA. TOMMA efficiently models and exploits a Markovian agent in a two player repeated game setting. Again we assume prior knowledge of a set of possible features (target set of features) some of which are assumed to characterize the unknown policy of the Markovian agent.

As our first test-bed for validating TOMMA's effectiveness , we introduce a challenging new domain - *The Surveillance Game*. The game is derived from the multi-robot patrol problem, a well-studied problem in the robotics community, e.g. [1, 53]. As our second test-bed, we focus on *The Ticket Checking* problem [42] which is inspired by a real life problem of catching passengers who do not buy a ticket (or evaders) while traveling on trains. For both of these domains, we pose the problem as learning in the presence of a feature based adversary and show how TOMMA models and exploits patterns in the adversary's behavior.

This contribution is presented in Chapter 6.

5. The fifth and final contribution of this book shows how our approach of modeling Markovian agents generalizes to solving a broader class of problems pertaining to the single agent RL setting, called "structure learning in factored state MDPs (FMDPs)" [4, 30]. Structure learning is the problem of learning the unknown structures in the underlying transition function of the FMDP from as few samples of online data as possible. Leveraging from our approach of modeling Markovian agents, we propose an alternative mechanism [21] of solving the structure learning problem that results

in sample complexity bounds which compare favorably with those provided by the existing approaches. We also empirically show that our approach competes well with the current state-of-the-art structure learning algorithm in certain representative benchmark domains. This contribution is presented in Chapter 7.

The remainder of the book is organized as follows. Chapter 2 presents the background necessary to understand all of the technical contributions of this book. Chapters 3 to 7 present all of our contributions as listed above. Chapter 8 presents some related work and situates our research. Finally Chapter 9 concludes by citing some possible avenues of future research along with some final remarks.

Chapter 2
Background

Our objective is to devise learning algorithms that an autonomous learning agent can use while functioning in a multiagent environment. In order to achieve this objective, we pose the learning problem as a Reinforcement Learning problem in a multiagent environment. Keeping in line with past research, we use the canonical game theoretic framework of repeated matrix games as our chosen multiagent environment.

This chapter serves two purposes. First, it reviews the concepts from Reinforcement Learning and repeated matrix games that are necessary for fully understanding the technical details of this book. Second, it establishes the notation that we use throughout this book. We begin in Section 2.1 by introducing the concepts related to Reinforcement Learning. Then in Section 2.2 we present the concepts related to repeated matrix games.

2.1 Concepts Pertaining to Reinforcement Learning

In the Reinforcement Learning (RL) framework, an agent makes its decisions as a function of a state signal which it receives from the environment. A state signal can be of many forms, one that remembers the entire history of past sensations to one that remembers nothing. Ideally we prefer a state signal that summarizes past sensations compactly, yet in such a way that all relevant information is retained. Such a state signal that succeeds in retaining all relevant information is said to be Markov, or to have the Markov property (we define this formally below).

For example, a checkers position (the current configuration of all the pieces on the board) would serve as a Markov state because it summarizes everything important about the complete sequence of positions that led to it.

If an environment has the Markov property, then its one-step dynamics enable us to predict the next state and expected next reward given the current state and action. An RL task that satisfies the Markov property is called a Markov Decision Process, or MDP [60]. If the state and action spaces are finite, then it is called a finite Markov Decision Process (finite MDP).

D. Chakraborty, *Sample Efficient Multiagent Learning in the Presence of Markovian Agents*, Studies in Computational Intelligence 523, DOI: 10.1007/978-3-319-02606-0_2, © Springer International Publishing Switzerland 2014

Definition 1. *Finite Markov Decision Process: A finite Markov Decision Process is given by a tuple $\{S, A, T, R\}$ where S is a finite set of states, A is a finite set of actions available to a learner, $T : S \times A \times S' \mapsto [0, 1]$ is the transition function and $R : S \times A \mapsto \Re$ is the bounded reward function. On taking an action $a \in A$ in state $s \in S$, the probability of transitioning to state $s' \in S$ and the reward obtained are given by $T(s, a, s')$ and $R(s, a)$ respectively.*

A RL agent finds itself in a MDP and it has to choose actions in different states to maximize the sum of rewards obtained. For that it needs to follow a policy: a strategy of choosing actions at different states. The better the policy, the higher is the sum of rewards obtained.

Definition 2. *Policy: A policy is a stationary strategy for choosing an action from a state in a MDP. A stationary strategy is one that produces an action based on only the current state, ignoring the rest of the agent's history. It can be formally defined as a function from states to actions; $\pi : S \mapsto A$.*

Let $U_T^\pi(s)$ and $U^\pi(s)$ be the T-step expected return and the infinite-horizon return respectively, from following a policy π when starting in state s in a MDP. More formally let $\mathbf{r_t}$ denote the expected reward at time step t from following π. Then,

$$U_T^\pi(s) = \frac{\sum_{t=0}^{T} \mathbf{r_t}}{T} \text{ and } U^\pi(s) = lim_{T \to \infty} U_T^\pi(s) \tag{2.1}$$

Following a policy in a MDP induces a Markov chain [60] on visited states. For simplicity of analysis, we assume that for any fixed policy of a MDP, the induced Markov chain is unichain [60]. We call such MDPs *unichain MDPs* [47]. Before, we formally introduce the concept of unichain MDPs, we need to introduce certain technical terminologies pertaining to MDPs.

Two states s_1 and s_2 *communicate* under a policy π if there is a positive probability of reaching each state from the other. A state is *recurrent* under a policy π if starting from that state, the probability of reaching that state again is non-zero. A non-recurrent state is called *transient*. A *recurrent class* of states is a set of recurrent states which also communicate with each other. A Markov chain is unichain if it has a single recurrent class of states along with a set of (possibly empty) transient states. Then a unichain MDP is formally defined as follows:

Definition 3. *Unichain MDP: A MDP is termed unichain if every policy in the MDP induces a unichain Markov chain, irrespective of the start state.*

For example Figure 2.1(a) shows a unichain Markov chain induced by following a specific policy in a MDP with two states s_1 and s_2. The policy is to follow action a_1 in both the states. Given that policy, the probability of transitioning from s_1 to s_2 and the reward associated with that transition are

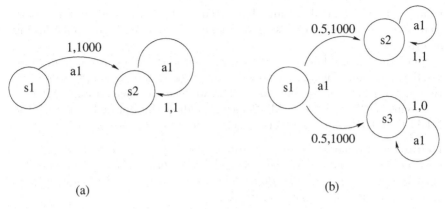

(a)   (b)

Fig. 2.1 Examples of a unichain and a multichain MDP

1 and 1000 respectively (denoted by 1,1000 in the figure). Similarly the probability of remaining in s_2 and the associated reward are 1 and 1 respectively. It is a unichain Markov chain as it has only one recurrent class of states $\{s_2\}$ and one set of transient states $\{s_1\}$.

On the contrary, for a *multichain* MDP, there exists at least one policy which leads to at least two recurrent classes of states along with a set of (possibly empty) transient states. For example Figure 2.1(b) shows a multichain Markov chain induced by following a specific policy in a MDP with three states s_1, s_2 and s_3. The policy is to follow action a_1 in all the states. Given that policy, the probability of transitioning from s_1 to either of s_2 or s_3 and the reward associated with that transition are 0.5 and 1000 respectively. The probability of remaining in s_2 and the associated reward are 1 and 1 respectively. The probability of remaining in s_3 and the associated reward are 1 and 0 respectively. It is a multichain Markov chain as it has two recurrent classes of states, namely $\{s_2\}$ and $\{s_3\}$, and one set of transient states $\{s_1\}$.

There are a number of interesting properties of unichain MDPs. Foremost among them is that for any policy, the infinite-horizon return from following that policy is independent of the start state and is a unique value. That is to say that the limit in Equation 2.1 exists for all s and is a unique value. For example in the unichain Markov chain from Figure 2.1(a), the infinite-horizon return from both the states is 1. Whereas in the multichain Markov chain from Figure 2.1(b), the infinite-horizon return from states s_1, s_2 and s_3 are 0.5, 1 and 0 respectively. Restricting our attention to just unichain MDPs simplifies our analysis while trying to compute an optimal policy for the MDP (optimal policy defined below), as we do not need to worry about different returns originating from different states. Note this is not a limitation of our approach, but just a simplifying assumption for the sake of analysis.[1] Our

---

[1] A very common assumption in RL literature while dealing with MDPs in average reward setting [47, 15, 52].

results naturally extend to multichain MDPs with just a small and necessary change to the definition of best performance we can expect from our learning algorithms.

Since the infinite-horizon return from all the states for a unichain finite MDP is the same for a fixed policy π, we denote it by a unique value U^π. Henceforth whenever we refer to a MDP, we mean a unichain finite MDP.

Solving a RL task for a MDP means finding the optimal policy of the MDP. For MDPs, we can precisely define an *optimal policy* as follows.

Definition 4. *Optimal policy: A policy that achieves the highest infinite-horizon return amongst all possible policies is the optimal policy. Let the optimal policy be π^* and its infinite-horizon return be U^*. Then formally U^* satisfies the following condition: $\forall \pi \neq \pi^*, U^* \geq U^\pi$.*

There are numerous algorithms for computing the optimal policy of a MDP. Instead of presenting them in detail, we point the reader to [52] for an excellent survey of these methods. For the purposes of understanding the technical details of our algorithms, an in-depth knowledge of these algorithms is not necessary. The knowledge of the fact that they do exist and can successfully compute an optimal policy for a MDP is sufficient.

A special class of MDPs are factored state MDPs (FMDP for short) [36] which exhibit a special structure in their state space and transition function. To be more precise, they assume that both the state space and the transition function can be factored into separate individual components.

Definition 5. *Factored state MDP: A factored state MDP, is a finite MDP where each state consists of n discrete factors. Formally $S = \{X_1, \ldots, X_i, \ldots X_n\}$ where each X_i assumes a value in the discrete range $D(X_i)$, called the domain of X_i. Furthermore the transition function of the FMDP is also factored and satisfies the conditional independence assumption. Let $s_{t+1}(i)$ be the value of factor X_i in state s_{t+1}. Let $T_{i,a_t}(s_t, s_{t+1}(i))$ be the probability of transitioning to $s_{t+1}(i)$ when action a_t is taken in state s_t. Then the conditional independence assumption implies that the following holds: $T(s_t, a_t, s_{t+1}) = \prod_{i=1}^n T_{i,a_t}(s_t, s_{t+1}(i))$.*

Since FMDPs are instances of MDPs, all concepts pertaining to MDPs (namely the concept of policy, optimal policy and return) extend naturally to FMDPs. For all of our contributions apart from contribution 5 (presented in Chapter 7), the setting is the general MDP.

While seeking theoretical guarantees about the quality of the time averaged return of a learning algorithm in a MDP after a finite number of steps, we need to take into account some notion of the mixing times of policies in the MDP. More formally, we need to understand the concept of the ϵ-return mixing time [47] of a policy in a MDP.

The standard notion of the ϵ-*mixing time* for a policy in a MDP quantifies the smallest number T of steps required to ensure that the distribution of visited states after T steps when following the policy is within ϵ of the stationary distribution induced by that policy where the distance between the distributions is measured by max norm or some standard measure. ϵ is generally a small value between 0 and 1. In contrast to the ϵ-mixing time, the ϵ-return mixing time only requires the expected return after T steps to approach the infinite-horizon return. The ϵ-return mixing time of π is defined as follows.

Definition 6. *ϵ-return mixing time: For an $0 < \epsilon < 1$, the ϵ-return mixing time T of a policy π is the smallest T such that $\forall T' \geq T$ and $\forall s$, $|U_{T'}^{\pi}(s) - U^{\pi}|_{\infty} \leq \epsilon$.*[2]

In other words, once we have executed a policy π for at least T steps where T is the ϵ-return mixing time of π, the expected return is always within a bound ϵ of U^{π}, irrespective of the start state.

This concludes our introduction of concepts pertaining to RL. We now proceed to introduce the concepts pertaining to our chosen game theoretic setting, namely repeated matrix games.

2.2   Concepts Pertaining to Repeated Matrix Games

Game theory [56] is a method of studying strategic decision making. More formally, it is the study of mathematical models of conflict and cooperation between intelligent rational decision makers. Game theory has been extensively used over the years in diverse fields such as economics, political science, logic and psychology. As the setting for our research, we focus on matrix games: the simplest and most well-studied of all game theoretic frameworks.

Definition 7. *Matrix game: A matrix game represents a scenario in which n agents are interacting with each other by simultaneously selecting actions. Without loss of generality, we assume that the set of actions available to all the agents are the same, i.e., $A_1 = \ldots = A_i = \cdots = A_n = A$. The payoff received by an agent i in the interaction is determined by a utility function over the agents' joint-action, $u_i : A^n \mapsto \Re$.*

Definition 8. *Bimatrix game: A bimatrix game is a special case of a matrix game with just two agents.*

Table 2.1 presents a couple of examples of famous bimatrix games with two actions for each agent. An *outcome* is a set of payoffs for all agents achieved as a result of a joint-action. Thus in Prisoner's Dilemma when both the agents play "cooperate", the resultant outcome is $(3, 3)$.

Definition 9. *Repeated game: A repeated game is a setting in which the agents play a matrix game repeatedly. While playing a repeated game each agent follows a policy to choose its action on each step. The general notion*

---

[2] $\|_{\infty}$ is the max norm.

Table 2.1 Payoff matrices for Prisoner's Dilemma (PD) and Chicken

(a) PD   (b) Chicken

	cooperate	defect
cooperate	(3,3)	(1,4)
defect	(4,1)	(2,2)

	swerve	bully
swerve	(3,3)	(2,4)
bully	(4,2)	(1,1)

of a policy for an agent in a repeated game is a function mapping each possible history of play to a distribution over its actions (a.k.a mixed action). Formally it is defined as follows: $\forall k \geq 0, \pi : A^{nk} \mapsto \Delta A$.[3] *An agent also achieves an expected return from playing the repeated game as defined in Definition 10.*

We say an agent i is playing a *stationary* policy if it plays the same mixed action at every time step. Note, the concept of stationary policy for a repeated game is different from the one pertaining to MDPs (Definition 2). The former is completely stateless while the latter implies playing the same action from a given state. Though we use the same terminology for both of these scenarios, the one we refer to would be obvious from the context. A stationary policy is called *pure* if it plays the same fixed action at every time step; otherwise it is called *mixed*. The set of policies for all agents is called a *joint-policy*.

One of our prime objectives is to pose the problem of learning the best response while playing an agent in a repeated game, which is either memory-bounded or Markovian, as learning an optimal policy in a MDP. In such scenarios, the utility function of the matrix game serves as the reward function of the MDP. Whereas the unknown policy of the other agent determines the state space and the transition function of the MDP. We make these connections more explicit when we introduce the concept of Adversary Induced MDP in Definition 17.

Definition 10. *Expected return: In a repeated game, when all the other agents follow their own share of a joint-policy, an agent i by following its own share of the joint-policy for T steps, denoted by π_i, achieves an expected return*

$$U_T^{\pi_i} = \frac{\sum_{t=1}^{T} \mathbf{r_t}}{T}$$

over those T steps. $\mathbf{r_t}$ is i's expected payoff at time t from following π_i given that all the other agents are following their own share of the joint-policy.

Definition 11. *Joint-return and social welfare: At any particular time step, the set of individual expected returns for all the agents playing a repeated game is called the joint-return on that time step. The sum of the elements of a joint-return is the social welfare (SW) value of that joint-return. More formally at any time step t, the set $\{U_t^{\pi_i}\}$ is the joint-return while $\sum_i U_t^{\pi_i}$ is the social welfare.*

---

[3] ΔA means a distribution over A.

For example in the case of the Prisoner's Dilemma (Table 2.1(a)), when both the agents keep playing "cooperate", the resultant joint-return at any particular time step has a social welfare value of 6.

A very crucial solution concept pertaining to learning in repeated games is the Nash equilibrium (named after John Forbes Nash, who proposed it) [55]. It is a joint-policy where no agent gains by unilaterally deviating to follow a different policy. In other words, if each agent chooses a policy and no agent benefits by changing its own policy unilaterally, then the corresponding joint-policy constitutes a Nash equilibrium joint-policy.

The most popular form of Nash equilibrium is the single stage Nash equilibrium. It is a stationary joint-policy that serves as a Nash equilibrium of both the single stage (the matrix game played just once) and the repeated game (when played repeatedly in every stage). It is a stationary joint-policy because each agent's policy is independent of the history of interactions so far and fixed for every time step. Formally a single stage Nash equilibrium is defined as follows. Let the set of all possible stationary policies for i be Π_i, while that of the other agents be Π_{-i}.

Definition 12. *Single stage Nash equilibrium (NE): Assume that all agents are following a stationary joint-policy. Let agent i's share of the stationary joint-policy be π_i while the rest of the agents' share be π_{-i}. Let U^{π_i} be i's expected payoff (utility) from following π_i when all the other agents follow π_{-i}, i.e., $U^{\pi_i} = \underset{a_i \sim \pi_i, a_{-i} \sim \pi_{-i}}{\mathbb{E}} (u_i(a_i, a_{-i}))$. We call this stationary joint-policy a single stage Nash equilibrium if for all such i's, the following inequality holds: $\forall \pi_i' \in \Pi_i, \pi_i' \neq \pi_i, U^{\pi_i'} \leq U^{\pi_i}$.*

For example in Prisoners Dilemma (Table 2.1(a)), the single stage NE for each agent is to play "defect". Once an agent plays "defect", there is no incentive for the other agent to deviate from playing "defect". Similarly the single stage NE in Rock-Paper-Scissors (R-P-S) (Table 2.2) is for each agent to play each action with probability $1/3$. Henceforth whenever we refer to a NE joint-policy, we mean the single stage NE joint-policy.

Often NE is a very hard solution concept to achieve primarily because of the difficulty of computing one for arbitrary matrix games. In such scenarios, we are often concerned with what the agent can achieve on its own as the best of all worst case scenarios. That leads us to the concept of security value for an agent in a matrix game. A security value (aka *maximin value*) for an agent is the maximum expected payoff it can guarantee on every time step regardless of the policies the other agents use.

Definition 13. *Security value: The security value SV_i for an agent i is the expected payoff it can guarantee on every time step regardless of the policies the other agents use. Formally it is defined as follows:*

$$SV_i = \max_{\pi_i \in \Pi_i} \min_{\pi_{-i} \in \Pi_{-i}} \underset{a_i \sim \pi_i, a_{-i} \sim \pi_{-i}}{\mathbb{E}} (u_i(a_i, a_{-i})) \qquad (2.2)$$

Table 2.2 Payoff matrix for Rock-Paper-Scissors

	rock	paper	scissor
rock	(0, 0)	(-1,1)	(1,-1)
paper	(1,-1)	(0, 0)	(-1,1)
scissor	(-1,1)	(1,-1)	(0, 0)

A stationary policy that guarantees the security value is called the *safety* policy. A safety policy for an agent can be computed through a simple linear program by solving Equation 2.2. For example in R-P-S (Table 2.2), playing each action with probability 1/3, guarantees an expected return of 0 to an agent, regardless of the policy the other agent uses.

As a starting point achieving the security value is a reasonable solution concept, but often a better return is achievable, especially when the other agent(s) exhibit limitations that can be modeled and exploited.

In practice, the other agent(s) may have unknown policies and may themselves be adapting. Ideally, we would like to develop algorithms that are guaranteed to perform "optimally" (yield maximal expected return) against *any possible* set of agent policies. However the prospect of doing so is limited by a variant of the No Free Lunch theorem [75]: any algorithm that tries to maximally exploit some class of agent policies can itself be exploited by some other class.

However, if one is willing to restrict the class of possible other agents to some finite set of policies, it is possible to develop learning algorithms that are guaranteed to perform well against agents drawn from this set. This book is concerned with modeling two such classes of agents: memory-bounded and Markovian. The class of Markovian agents subsume the class of memory-bounded agents. We formally introduce these two classes of agent policies next. But before we do that we introduce the concept of a bounded history.

Definition 14. *Bounded history: A bounded history is a vector of length K consisting of the past K joint-actions played by the agents. Formally we denote it by h^K where K is the window size or length of the joint history. $h^K(j)$ is the jth joint-action in the sequence h^K where $0 \leq j < K$ with $h^K(0)$ being the most recent joint action. Similarly $h_i^K(j)$ is the jth action played by agent i in the sequence h^K with $h_i^K(0)$ being the most recent action played by i. The bounded history at time t is denoted by $h^{K,t}$.*

We formally denote the set of all feasible bounded histories of size K as $\mathcal{H_K}$. Note, while playing against a set of memory-bounded agents of memory size K, not all bounded histories of size K are reachable. For example while playing against an agent that never plays a specific action, it is impossible to have any bounded history which has that agent playing that specific action.

Definition 15. *Memory-bounded agent: A memory-bounded agent characterized by its memory size K chooses its next mixed action as a function of the*

most recent K joint-actions played in the current history. Formally its policy
π *is defined as* $\pi : \mathcal{H}_{\mathcal{K}} \mapsto \Delta A$.

Memory-bounded agents occur frequently in the literature of repeated
games. For example the famous *tit-for-tat* policy [56] for playing repeated
Prisoner's Dilemma which leads to two rational agents coordinating by playing "cooperate", is a memory-bounded policy with memory size 1. The agent
only remembers the last action played by the other agent and repeats that
in the current time step.

Definition 16. *Markovian agent: A Markovian agent decides on its next step
action by consulting the values of a set of features, denoted by F. A feature
$f \in F$ is a finite discrete valued statistic computed from the joint history of
play. Each such feature transitions in a fashion such that its value at time $t+1$
depend only on the collective values of all the features and the joint-action
taken, both from time t (Markov property [60]).*

Note that memory-bounded agents are a special case of Markovian agents
whose feature space is the set of joint-actions from a bounded history of
play. Below we present more examples. Henceforth for the introduction of
the remaining concepts, we assume that there are just two agents playing the
repeated game, where one of the agents is under our control and denoted by
i. The other agent and its unknown policy are o and π_o respectively.

The setting is repeated R-P-S. Assume o plays each action randomly unless
i has chosen the same action on the last 5 consecutive plays, in which case it
plays the best response to that action. In this case π_o can be represented as
a function of the past 5 actions played. But it can also be represented more
efficiently with just two features, namely the last action played by i and how
many consecutive times (up to 5) that action has been played. o is Markovian
because the next step values of these two features depend only on their most
recent values and the last action selected by i.

However, the concept of Markovian agents is far more general than plain
memory-bounded agents and captures agents whose policies are based on the
entire history of play. Consider the following example of two agents playing
the coordination game from Table 2.3. Assume o follows a coordination policy
which is as follows. If the number of times i played action "Heads" over the
entire history of play is even, then play "Heads", else play "Tails". π_o can
easily be modeled with just one boolean feature, namely whether the number
of times i played "Heads" over the entire history of play is even or not. o is
Markovian because the next step value of this feature depends only on its
most recent value and the last action selected by i.

The key insight enabling our research is that in a scenario where o is
either memory-bounded or Markovian, the dynamics of playing against o
can be modeled as a MDP. We illustrate this with the example of o being a
memory-bounded agent with memory size K. Similar analogies can be drawn
for the case when o is Markovian. Then the dynamics of playing against o can
be modeled as a MDP whose transition probabilities and reward function are

Table 2.3 Payoff matrix for coordination game

	Heads	Tails
Heads	(1, 1)	(0,0)
Tails	(0,0)	(1, 1)

determined by π_o. That is for a bounded history of play (a "state") $h^{K,t}$, the next state $h^{K,t+1}$ and the reward received are determined by the current state $h^{K,t}$, o's policy for that specific state $\pi_o(h^{K,t})$, and the action a_i chosen by agent i at time step t. We call such a MDP an Adversary Induced MDP [8, 19].

Definition 17. *Adversary Induced Markov decision process (AIM): An Adversary Induced MDP characterized by a memory size K is defined as follows,*

State Space: The state space is $\mathcal{H}_K$; the set of all feasible bounded histories of length K.

Action Space: The action space is A; the set of actions available to agent i.

Transition Function: The bounded history (state) is updated as a sliding window. Transitioning from a state $h^{K,t}$ to a state $h^{K,t+1}$ is just keeping the last $K - 1$ joint-actions (each shifted one time step backwards) and including the latest pair at index 0 of the vector. Let the latest joint-action in $h^{K,t+1}$ be (a_i, a_o). Then the probability with which this transition happened is $\pi_o(h^{K,t}, a_o)$: the probability with which o plays action a_o given the current state is $h^{K,t}$.

Reward Function: The reward obtained by i when it takes action a_i in state $h^{K,t}$ and o plays an action $a_o \sim \pi_o(h^{K,t})$ ('$\sim$' means drawn based on $\pi_o(h^{K,t})$) is $u_i(a_i, a_o)$.

Consider the following example of an AIM. In R-P-S assume o is a memory-bounded with $K = 1$. Let the current state be (R, P), meaning that on the previous step, agent i selected R and o selected P. Assume that from that state, o's policy is to play actions R, P and S with probability $0.25, 0.25$, and 0.5 respectively. When i chooses to take action S in state (R, P), the probabilities of transitioning to states $(S, R), (S, P)$ and (S, S) are then 0.25, 0.25 and 0.5 respectively. Transitions to states that have a different action for i, such as (R, R), have probability 0. The reward obtained by i when it transitions to state $(S, R), (S, P)$ and (S, S) are -1, 1 and 0 respectively. Thus, both the transition and the reward functions follow the Markovian dynamics and are completely determined by the unknown π_o.

By modeling the interaction dynamics as an AIM, we can find the optimal policy of playing against o (best response) by solving for the optimal policy of the AIM. If π_o were known, then we could compute the optimal policy via Dynamic Programming (using techniques such as Value Iteration). However, since we assume that π_o is unknown, we need to solve for the optimal policy of the MDP using online Reinforcement Learning methods: a key goal of the learning algorithms presented in this book.

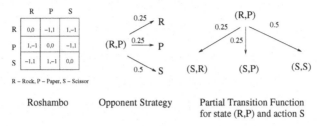

| Roshambo | Opponent Strategy | Partial Transition Function for state (R,P) and action S |

Fig. 2.2 Example of the partial transition function for state (R,P)

As suggested earlier, we assume that the induced AIM is a unichain MDP. In that case, we need not worry about different returns originating from different states while following a policy. Also, it is important to realize that there exist Markovian agents which do not allow convergence to the optimal policy of the induced MDP once a certain set of moves has been played. For example, the *grim-trigger* agent in Prisoner's Dilemma plays "cooperate" at first, but then plays "defect" forever once the other agent plays "defect" even once. Grim-trigger is Markovian because its policy is based on only one Markovian boolean feature: whether the other agent ever played "defect". There is no way of detecting its policy without defecting, after which it is impossible to recover to the optimal policy that leads to mutual cooperation. Thus, in our analysis we constrain the class of Markovian agents to include only those which do not negate the possibility of convergence to optimal exploitation, given any initial sequence of exploratory moves.

That concludes our introduction of concepts pertaining to RL and repeated matrix games. The above alluded concepts are sufficient for understanding most of the technical contributions of this book. We reserve a small amount of notation of local relevance for later chapters.

Chapter 3
Learn or Exploit in Adversary Induced Markov Decision Processes

Our overarching goal is to model agent policies which are common in day to day life and are exploitable. A natural form of such an agent policy is a memory-bounded one where the agent only remembers a finite number of past samples of data to decide on its current action.

There are three reasons which motivate us to consider a memory-bounded agent as a candidate agent behavior for our learning algorithms to model and exploit. First, memory-bounded behavior is quite prevalent in day to day life. For example, often while deciding whether we should visit a restaurant or watch a movie pertaining to a particular director, our decision is guided by our most recent experiences from having performed that action. Second, in practice every agent has a finite memory. For example if an agent is a computer, its memory is limited by its primary and secondary storage capacity. Third and most importantly, despite how restrictive it might appear, a large set of agents from both the game theory and MAL literature are in fact memory-bounded. Common examples include Godfather [68], polynomial Nash policy [51] and Bounded Fictitious Play [61]. Furthermore, if we consider agents whose future behavior depends on the entire history, we lose the ability to (provably) learn anything about them in a single repeated game, since we see a given history only once. The concept of memory-boundedness limits the agent's ability to condition on history, thereby giving us a chance to learning its policy online.

Motivated by these observations, there is a large amount of literature that attempts to model memory-bounded agents [8, 58, 59]. This chapter introduces a new algorithm that adds to that list called *Learn or Exploit in Adversary Induced Markov Decision Processes* (or LoE-AIM for short) that models any agent that can be treated as a memory-bounded agent with a fixed memory size, in a bimatrix game. LoE-AIM makes no prior assumptions about the other agent apart from the fact that it is memory-bounded with a fixed memory size and its memory size is known. We relax this assumption in Chapter 4 where we present a learning algorithm that is unaware of the exact memory size of the other agent, but only a very loose

D. Chakraborty, *Sample Efficient Multiagent Learning in the Presence of Markovian Agents*, Studies in Computational Intelligence 523, DOI: 10.1007/978-3-319-02606-0_3, © Springer International Publishing Switzerland 2014

upper-bound of it. In repeated play, LoE-AIM either explores and gathers new information about the other agent or converges to the best response to its partially learned policy. LoE-AIM is fully implemented with results in this chapter demonstrating its superiority over representative algorithms from the literature.

Though memory-boundedness is a stricter assumption than we ultimately want, we relax this assumption in Chapter 5 and onwards where we deal with modeling Markovian agents. Focusing first on this restrictive case allows us to lay the framework for the algorithms that follow later. The chapter is organized as follows: Section 3.1 introduces LoE-AIM, Sections 3.2 and 3.3 presents empirical results demonstrating the performance of LoE-AIM in representative bimatrix games and against representative agents, and Section 3.4 summarizes the chapter.

3.1 LoE-AIM

Our focus is on modeling memory-bounded agents as ones which induce a Markov Decision Process (MDP) according to the Adversary Induced MDP (AIM) model (See Definition 17 from Chapter 2). By this model the learner finds itself in a MDP whose state space is the set of all feasible bounded histories and whose transition and reward functions are determined by the other agent's policy. This section introduces LoE-AIM, our learning algorithm for the AIM setting. We present two versions of LoE-AIM: one tailored to model memory-bounded agents which play deterministically and the other for memory-bounded agents which play stochastically. The former can be used if we are certain that the other agent is deterministic. If not, the latter should be used. In both the versions, we assume that K is the true memory size of the other agent and LoE-AIM is aware of it. Henceforth we denote the other memory-bounded agent as o and its policy as π_o (see Definition 15 from Chapter 2), while the agent under our control as i.

Algorithm 1 presents the version of LoE-AIM for modeling deterministic agents. The algorithm takes as input the current model of π_o (denoted as $\hat{\pi}_o$), the current start state of the AIM (denoted as "s" in Algorithm 1; initial K sized bounded history) and the number of episodes for which it should continue learning. To begin with, $\hat{\pi}_o$ refers to some partially learned model if it exists. If the algorithm has no prior information about o, $\hat{\pi}_o$ comprised of zero mappings. All the results presented in this chapter assume that there exists no such partial model and LoE-AIM learns from scratch.

Since π_o is deterministic, just one visit is needed to a state of the AIM to know what is o's policy for that state. The SOLVE-AIM function finds a control policy (π_i) by solving for the optimal policy of the AIM assuming that for every known state s of play, o plays $\hat{\pi}_o(s)$ and for all unknown states, o plays the *maximax policy* for i (a friendly policy that maximizes the maximum payoff for i).

Algorithm 1. LoE-AIM-DETERMINISTIC

---

begin

    input : $\hat{\pi}_o$, s, episodes

    ;

    episode $\leftarrow$ 0;

    $\pi_i \leftarrow$ SOLVE-AIM($\hat{\pi}_o$);

    repeat

        $a_o \leftarrow$ action taken by agent;

        $a_i \leftarrow$ action as per π_i;

        if *s not visited before* **then**

            $\hat{\pi}_o \leftarrow$ UPDATE-MODEL($\hat{\pi}_o$, s, a_o);

            $\pi_i \leftarrow$ SOLVE-AIM($\hat{\pi}_o$);

            s $\leftarrow$ UPDATE-HISTORY(s, a_i, a_o);

            episode $\leftarrow$ episode + 1;

    until *episode < episodes*;

For example in PD (Table 3.1(a)), the maximax policy for o is to play "cooperate". i can then play "defect" and achieve the maximum payoff of 4. Similarly in Chicken (Table 3.1(b)), the maximax policy for o is to play "swerve". i can then play "bully" and achieve the maximum payoff of 4. In Rock-Paper-Scissors, the maximax policy can be playing any of the actions with probability 1. There always exists an action that beats the chosen action resulting in the maximum payoff of 1.

This *optimism under uncertainty* assumption for unknown states forces π_i to explore states not visited before. Whenever a new state s is visited for the first time, the UPDATE-MODEL method updates the current model $\hat{\pi}_o$ with the new information: what o plays for s. The UPDATE-HISTORY method updates s by prepending the most recent joint-action and removing the oldest joint-action.

Algorithm 2 is similar to Algorithm 1 except now π_o can be stochastic. In this case, i maintains a stochastic model of π_o. The only structural difference from Algorithm 1 is when and how the model $\hat{\pi}_o$ is updated (Line 6 of Algorithm 2). Algorithm 2 updates the model when any state s gets visited for the m'th time. The updated model now captures o's policy for state s (i.e., $\hat{\pi}_o(s)$) which is just the empirical distribution of o's play for state s based on the first m samples of data, i.e., first m visits to s. All results in this chapter use $m = 10$, the value that led to the best results in informal preliminary testing.

It follows that in repeated play LoE-AIM either converges to the optimal policy for the partially learned agent model or keeps expanding the learned model. Let $\bar{\pi}_o$ be the remainder of π_o that needs to be learned at a particular time instant, while $\hat{\pi}_o$ refers to the part of π_o that i knows. By solving for a control policy where for every state in $\bar{\pi}_o$, i believes that it could get the best possible reward (since it assumes that o plays the maximax policy for i

---

Algorithm 2. LoE-AIM-STOCHASTIC

---

begin
  input : $\hat{\pi}_o$, s, episodes
  ;
  episode $\leftarrow$ 0;
  π_i $\leftarrow$ SOLVE-AIM($\hat{\pi}_o$);
  repeat
      a_o $\leftarrow$ action taken by agent;
      a_i $\leftarrow$ action as per π_i;
      if s *visited* m *times* **then**
          $\hat{\pi}_o$ $\leftarrow$ UPDATE-MODEL($\hat{\pi}_o$, s, a_o);
          π_i $\leftarrow$ SOLVE-AIM($\hat{\pi}_o$);
          s $\leftarrow$ UPDATE-HISTORY(s, a_i, a_o);
          episode $\leftarrow$ episode + 1;
  until *episode < episodes*;

---

at those states and hence i gets the maximum reward of the matrix game for those states), the algorithm generates π_i that promotes exploring states in $\hat{\pi}_o$. However there exists a probability that LoE-AIM might get stuck in a sub-space of the AIM from where certain other states are not reachable (due to π_o), or reachable with a very low probability. In those cases, it converges to the best policy pertaining to the partially learned model. This exploratory aspect of LoE-AIM is motivated in part by the RMAX algorithm [15] which also deliberately balances exploitation with exploration of unvisited states.

3.2   Results against Deterministic Agents

This section presents the results achieved by LoE-AIM (deterministic version) against certain deterministic agents, namely k-Markov agents such as Godfather [68], Bully [68] and Best Response (BR). We begin by introducing these policies.

Godfather is a finite-state policy that makes the other agent i an offer that it cannot refuse. It strategizes as follows. First it chooses a targetable pair. A *targetable pair* is a stationary pure joint-policy which assures a payoff greater than the security value for each agent. We only focus on games where such a targetable pair exists. Then if i plays its half of the targetable pair on time step t, Godfather plays its half in time step $t + 1$. Otherwise, it plays the minimax policy (threat) that forces i to achieve at most its security value. So there remains no incentive for i not to play its own share of the targetable pair which ensures it a return greater than its security value. Godfather is a memory-bounded agent with memory size 1, i.e., $K = 1$. For example in the case of PD (Table 3.1), Godfather targets the { "cooperate", "cooperate" } pair and uses "defect" as the threat policy. We now introduce a couple of variations of the Godfather policy that are tailored for $K > 1$.

Table 3.1 Payoff matrices

(a) Prisoner's Dilemma		
	cooperate	defect
cooperate	(3,3)	(1,4)
defect	(4,1)	(2,2)

(b) Chicken		
	swerve	bully
swerve	(3,3)	(2,4)
bully	(4,2)	(1,1)

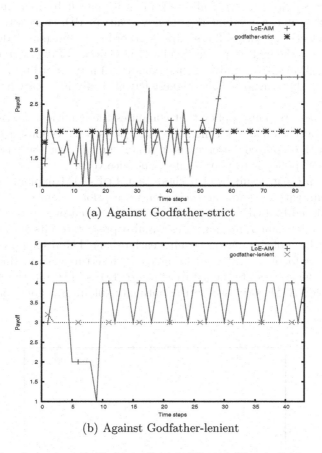

(a) Against Godfather-strict

(b) Against Godfather-lenient

Fig. 3.1 Results against Godfather policies in PD with $K = 3$. Against Godfather-strict, LoE-AIM eventually learns that it should play "cooperate" (its half of the targetable pair) and hence converges to a payoff of 3. For Godfather-lenient, LoE-AIM learns to optimally exploit by playing "cooperate" frequently enough so that the history always contains one "cooperate" action for i.

- Godfather-lenient plays its part of a targetable pair if i played its own half of the pair at least once within the last K time steps. Otherwise Godfather-lenient punishes i by playing the threat policy that reduces i's expected payoff to its security value.
- Godfather-strict is a stricter version that punishes i if i ever deviated from the targetable pair during the past K time steps.

Bully is a pure stationary policy given by $\text{argmax}_{a_o \in A} u_o(a_i^*, a_o)$ where $a_i^* = \text{argmax}_{a_i \in A} u_i(a_i, a_o)$. Bully optimizes its payoff by assuming that i is the follower and will adapt accordingly. Thus in PD, the Bully policy is to play "defect". i which follows adjusts accordingly and plays "defect", as that ensures it the highest payoff. Similarly in Chicken (Table 3.1), the Bully policy is to always play "bully". i which follows adjusts accordingly and plays "swerve", as that ensures it the highest payoff. Bully is a stationary policy with $K = 0$.

BR is a best response policy that computes the empirical distribution of i's actions over the past K time steps, and plays best response to it. So in Chicken, if the i played "swerve" more than "bully" over the past K time steps, the BR policy is to play "bully", and vice versa.

Now we present results which show that LoE-AIM exploits all of the above agents without knowing their identity a priori. Figure 3.1 shows the results achieved by LoE-AIM in PD against the variations of the Godfather policies. For benchmarking purposes we also present results from self-play (the Godfather policies playing against each other). The results presented in Figure 3.1 are for $K = 3$. We ran 30 runs of each setting with random instantiations of the start state, i.e., start state for LoE-AIM and the godfather policies. LoE-AIM converges to the optimal exploitation policy in each of the runs.

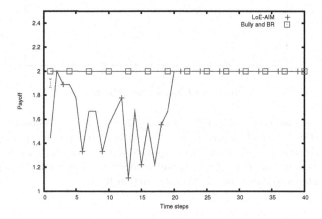

Fig. 3.2 Results against Bully and BR in PD with $K = 3$. Both the Bully and BR policy is to play "defect" always. Against both of these policies, LoE-AIM eventually learns to play "defect" and converges to a payoff of 2.

Against Godfather-strict, LoE-AIM eventually learns that it should play "cooperate" (its half of the targetable pair) and hence converges to a payoff of 3 (see Figure 3.1(a)). For Godfather-lenient, LoE-AIM learns to optimally exploit by playing "cooperate" frequently enough so that the history always contains one "cooperate" action for i. At convergence, the LoE-AIM player plays "defect" twice followed by a cooperate ensuring two consecutive payoffs of 4 followed by a payoff of 3 (Figure 3.1(b) shows the plot for one such run).

In PD, both the Bully and BR policy is to play "defect" always. Against both of these agents, LoE-AIM eventually learns to play "defect" and converges to a payoff of 2 (see Figure 3.2).

3.3   Results against Stochastic Agents

We now present results of LoE-AIM (stochastic version) learning against popular MAL algorithms that converge to a Nash equilibrium in self-play. Our results are against WOLF-IGA [13] and REDVALER [9]. We refer the reader to the respective citations for the details of these algorithms. An important point worth noting about these algorithms is that neither is memory-bounded and instead each uses the entire history of interactions to decide their current mixed action. However, our results show that even for $K = 4$, LoE-AIM can efficiently model these agents and exploit them optimally in certain games. Once again all our results are averaged over 30 runs. For LoE-AIM, the start state is chosen randomly for each of these runs. The learning rates used for WOLF-IGA are 0.08 and 0.04, while that for REDVALER is 0.05.

Figure 3.3 gives evidence that LoE-AIM was successful in reaching its optimal payoff in Chicken (Table 3.1(b)) by exploiting the MAL agents on both the occasions. The reason we choose Chicken is because the game has three Nash equilibria: two in pure policies, sustaining the outcomes (4,2) and (2,4), and one in mixed policies where the players play each of their actions with equal probability with the corresponding expected payoff of 2.5 for each agent. Neither WOLF-IGA, nor REDVALER guarantees the possible final converged Nash outcome in self-play, e.g, in both Figures 3.3(a) and 3.3(b), self-play between these MAL algorithms generates a payoff much less than 4 showing that on numerous occasions the final converged Nash outcome was not (4,2), the one most coveted by i. In contrast, in all of its runs, LoE-AIM converged to the outcome (4,2). The difference in the total payoff accrued by LoE-AIM and the MAL algorithms after 350 episodes of learning is statistically significant by a T-test (p-value < 0.05).

Finally Figure 3.5 shows an evaluation of head to head comparison amongst the various agent policies discussed in this chapter. There are 78 structurally distinct bimatrix games with two actions for each player in which the two players can strictly rank the four outcomes from best to worst (a ranking from 4 to 1). Of the 78 games, only 6 games (shown in Figure 3.4) have multiple Nash equilibria with each player favoring a different one. We present results

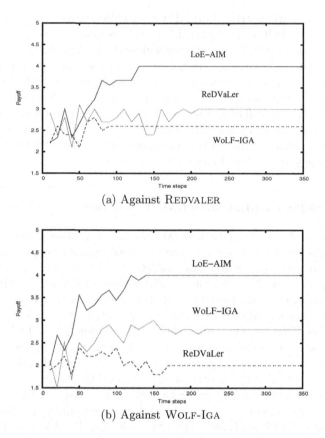

(a) Against REDVALER

(b) Against WOLF-IGA

Fig. 3.3 Results against REDVALER and WOLF-IGA in Chicken, $K = 4$. Neither WOLF-IGA, nor REDVALER guarantees the possible final converged Nash outcome in self-play, e.g, in both cases, self-play generates payoffs much less than 4 showing that on numerous occasions the final converged Nash outcome was not (4,2), the one most coveted by i. In contrast, in all of its runs, LoE-AIM converged to the outcome (4,2).

from these games because in self-play neither of the tested MAL algorithms (WOLF-IGA and REDVALER) guarantee the final converged Nash outcome. Each point in the plot has been averaged over results achieved from all the 6 games, with the results in each game averaged over 30 runs.

For benchmark comparisons, we show head to head results achieved by various other algorithms that i could have used as its default policy instead of LoE-AIM. Figure 3.5 shows that against the MAL algorithms (REDVALER and WOLF-IGA) and BR, LoE-AIM (stochastic version) successfully converges to its best payoff of 4 on all occasions thereby demonstrating its ability to exploit them optimally. All the benchmarks generate lower average payoffs

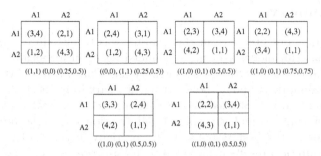

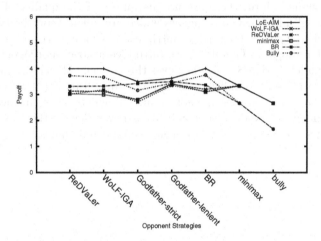

Fig. 3.4 Payoff matrices of 6 games with multiple Nash equilibria. ((1,1) (0,0) (0.25,0.5)) means that the game has 3 Nash equilibria where the probabilities of playing action A1 by both players are respectively (1,1), (0,0) and (0.25,0.5).

Fig. 3.5 Result against the 6 games with multiple Nash equilibria. Against the MAL algorithms (REDVALER and WOLF-IGA) and BR, LoE-AIM successfully converges to its best payoff of 4 on all occasions thereby demonstrating its ability to exploit them optimally. All the benchmarks generate lower average payoffs when played against these agents. Against the other agents, LoE-AIM does better than all the other benchmarks (though the average payoff is lower than 4).

when played against these agents. Against the other agents, LoE-AIM does better than all the other benchmarks though the average payoff is lower than 4 in these cases. Note, that against certain agents it is not possible to achieve the payoff of 4 because it won't allow that, e.g, against Godfather-strict in PD (see Figure 3.1(a)). The difference in the average payoff accrued by LoE-AIM and the other algorithms is statistically significant by a T-test (p-value < 0.05) in most occasions.

3.4 Summary

To summarize, our objective in this chapter was to introduce a simple algorithm that can model a memory-bounded agent in a repeated bimatrix game assuming it knows the agent's memory size. Called LoE-AIM, it makes no prior assumptions about the other agent apart from the fact that it is memory-bounded with a fixed known memory size. In repeated play, LoE-AIM either explores and gathers new information about the other agent or converges to the best response to its partially learned policy. We also presented empirical results demonstrating its superiority over representative algorithms from the literature.

However there are three shortcomings of LoE-AIM. First, it assumes that the true memory size of the agent is known beforehand. Second, the guarantee it provides mostly relies on empirical testing, without a proper theoretical backing. Third, there are no guarantees of the quality of return it achieves in self-play or against agents which are not memory-bounded with a known memory size. Our next contribution (Chapter 4) addresses all of these shortcomings. In Chapter 4 we introduce a novel multiagent learning algorithm called CMLeS which achieves three crucial objectives, namely (1) converges to a Nash equilibrium joint policy in self-play, (2) achieves close to the best response against memory-bounded agents whose memory size is upper-bounded by a known value and (3) achieves close to the security value against any other set of agents, in arbitrary repeated games.

Chapter 4
Convergence, Targeted Optimality and Safety in Multiagent Learning

In the previous chapter, we presented an algorithm LoE-AIM that models memory-bounded agents assuming that the memory size of these agents is known beforehand. In situations where such prior knowledge is unavailable, a possible solution can be to use a very large memory size that suffices to be a conservative upper-bound of the true unknown memory size. Lets call this conservative estimate K_{max}. Recall that in such a scenario, LoE-AIM in the worst case requires visits to all bounded histories of size K_{max} before it can converge to following an exploitive policy. Since the number of bounded histories of size K_{max} is exponential in K_{max}, this renders LoE-AIM highly inefficient (and impractical) for modeling agents with a big K_{max}. One of the main goals in this chapter is to address this shortcoming of LoE-AIM and propose an algorithm whose sample complexity scales exponentially with the true memory size, not with the conservative upper-bound K_{max}. This gives us the freedom to choose a loose K_{max} without having to worry about the sample complexity of achieving close to the best response being adversely affected by that.

In fact our overarching objective in this chapter is not just to solely focus on modeling memory-bounded agents. In the true spirit of MAL, we seek some guarantee against any possible set of agents playing the repeated game. In this regard, the contribution of this chapter (the first big contribution of this book) is a novel MAL algorithm called *C*onvergence with *M*odel *L*earning and *S*afety (or CMLeS for short)[1] that for a multi-player multi-action (arbitrary) repeated game, achieves the following three objectives (the first in the literature to do so):

Convergence [13]: converges to following a Nash equilibrium joint-policy in self-play (other agents are also CMLeS);

Targeted optimality [59]: achieves close to the best response with a high probability, against a set of memory-bounded agents whose memory size is upper-bounded by a known value K_{max};

---

[1] Pronounced as "seamless";

D. Chakraborty, *Sample Efficient Multiagent Learning in the Presence of Markovian Agents*, Studies in Computational Intelligence 523, DOI: 10.1007/978-3-319-02606-0_4, © Springer International Publishing Switzerland 2014

Safety [32]: achieves close to the security value against any other set of agents which cannot be represented as being K_{max} memory-bounded;

CMLeS improves upon the state-of-the-art MAL algorithm for modeling memory-bounded agents, called PCM(A) [59], in the following two ways.

1. The only guarantee of optimality against memory-bounded agents that PCM(A) provides is against the ones that are drawn from an initially chosen target set. In contrast, CMLeS can model any memory-bounded agent(s) whose memory size is loosely upper-bounded by K_{max}. Thus it does not require a target set of agents as input: its only input in this regard is K_{max};
2. Once convinced that the other agents are not self-play agents, PCM(A) achieves targeted optimality against memory-bounded agents by requiring that all feasible bounded histories of size K_{max} be visited a sufficient number of times. K_{max} for PCM(A) is the maximum memory size of any agent from its target set.

   For CMLeS, K_{max} serves to be a conservative upper-bound of the true memory size (say K). To achieve targeted optimality, requiring visits to all feasible bounded histories of size K_{max} may be very wasteful if K is significantly smaller than K_{max}. Our key theoretical result concerning CMLeS shows that it achieves targeted optimality by requiring a sufficient number of visits to only all feasible bounded histories of size K. In that way CMLeS is much more sample efficient than PCM(A).

We are going to introduce CMLeS in two parts. In the first part, we introduce a subroutine of CMLeS called MLeS that ensures targeted optimality against memory-bounded agents and safety against any other set of agents. Then we are going to build on MLeS to propose the fully blown CMLeS algorithm that achieves convergence as well. The remainder of the chapter is organized as follows. Section 4.1 presents MLeS, Section 4.2 presents the fully blown CMLeS algorithm, Section 4.3 presents some empirical results while Section 4.4 summarizes the chapter.

4.1 Model Learning with Safety (MLeS)

In this section, we introduce an algorithm, Model Learning with Safety (MLeS for short), a sub-routine of CMLeS that ensures targeted optimality against memory-bounded agents and safety against any other set of agents. Assume a repeated game between MLeS and a set of other agents. MLeS then achieves the following two objectives:

Targeted optimality: if these other agents are memory-bounded with their memory size upper-bounded by K_{max}, MLeS then achieves close to the best response with a high probability;

Safety: achieves close to the security value against any other set of agents which cannot be represented as being K_{max} memory-bounded;

We begin by showing how MLES achieves the targeted optimality objective. Let the MLES agent (the agent under our control) be i and the set of memory-bounded agents be a single agent o, with K and π_o being its unknown memory size and policy respectively.[2] We assume that $K_{max} \geq K$. Also to keep the analysis simple, we assume that the action space for o is just A.

We can always model π_o by assuming that it is memory-bounded with memory size K_{max}. A *model* for π_o is defined as follows.

Definition 18. *Model: A model $\hat{\pi}_k$ of π_o is defined by a possible memory size $k \leq K_{max}$ and specifies a distribution over the action set A (mixed action) for every feasible bounded history of size k, i.e., $\hat{\pi}_k : \mathcal{H}_k \mapsto \Delta A$.*

Note that modeling π_o based on K_{max} may involve learning over a much larger state space than is necessary. Our goal is to model π_o with the shortest most descriptive model (the model pertaining to the true memory size K or less).

MLES is introduced in Algorithm 3. For the sake of clarity, we break our algorithmic analysis of MLES into five parts. First in Section 4.1.1, we discuss the choice of the inputs for MLES. Second in Section 4.1.2, we describe how MLES operates from a high-level. Third and fourth in Sections 4.1.3 and 4.1.4 we focus on MLES's two main algorithmic components: the FIND-MODEL algorithm and its action selection mechanism respectively. Finally in Section 4.1.5, we remove a crucial assumption made in the aforementioned four sections (namely Assumption 1 from Section 4.1.1) and complete our specification of MLES.

4.1.1 Inputs to MLES

The inputs to MLES are ϵ, δ, T and K_{max}. Both ϵ and δ are small probability values. T is the planning horizon explained in the next paragraph. A reader not interested in a deep theoretical understanding of MLES may skip the rest of this subsection and treat these inputs as free parameters. We devote the rest of this subsection justifying the choice behind these input parameters that facilitates our theoretical claims concerning MLES.

MLES operates by planning for T time steps at a time. In each such planning iteration, it uses the best model of π_o at hand and plans its actions for the next T time steps based on it. Let U^* be the expected return from the best response against o, i.e., the optimal return achievable in the AIM induced by π_o. To facilitate the theory behind our claim that MLES converges to follow the best response against o, we assume that the (ϵ, T) pair taken as input always satisfies the following assumption:

---

[2] Note, a set of memory-bounded agents with a memory size K can be treated as a single memory-bounded agent of same memory size whose policy is just the joint-policy of all the agents.

Assumption 1. *The planning horizon T is sufficiently large and the ϵ sufficiently small to ensure that*

1. *T is the ϵ-return mixing time of the optimal policy for the AIM;*
2. *for any sub-optimal policy π and for any state s of the induced AIM, $U_T^\pi(s) < U^* - 2\epsilon$;*

Another way of thinking of Assumption 1 is that if we achieve a T-step expected return as high as $U^* - 2\epsilon$ in the underlying AIM from any start state, then we must be following the optimal policy for the AIM.

A pertinent question is whether for any set of memory-bounded agents such an (ϵ, T) pair exists or not. Let $\hat{U}$ be the expected return in the AIM from the best sub-optimal policy. Lets choose an ϵ smaller than $\frac{U^* - \hat{U}}{3}$. Let T be the maximum of all ϵ-return mixing times from all policies. Clearly this choice of an (ϵ, T) pair satisfies Assumption 1. Hence for any set of memory-bounded agents, there exists an (ϵ, T) pair that satisfies Assumption 1.

Our initial analysis caters to the special (and the more interesting) case where we assume that MLES is aware of such an (ϵ, T) pair that satisfies Assumption 1 (Sections 4.1.2 to 4.1.4). Later in Section 4.1.5, we show how a simple extension of our solution for this special case solves the more general problem where MLES is unaware of such an (ϵ, T) pair a-priori.

4.1.2 High Level Idea behind MLES

This subsection provides the high level idea behind MLES (Algorithm 3). Since MLES is unaware of the exact K that characterizes π_o, it maintains a model of π_o for every $0 \leq k \leq K_{max}$. Thus it maintains $K_{max} + 1$ models in total. Let the model that is based on the past k joint-actions be $\hat{\pi}_k$. Internally each $\hat{\pi}_k$ maintains a value $M_k(\mathbf{b_k})$ which is the maximum likelihood distribution of o's play, for every possible value $\mathbf{b_k}$ of the past k joint-actions. Whenever the past k joint-actions assume a value $\mathbf{b_k}$ in online play, we say a *visit* to $\mathbf{b_k}$ has occurred. $\hat{\pi}_k(\mathbf{b_k})$ is then defined as follows:

$$\hat{\pi}_k(\mathbf{b_k}) = \begin{cases} M_k(\mathbf{b_k}) & \text{once } visit(\mathbf{b_k}) = m_k \\ \perp & \text{when } visit(\mathbf{b_k}) < m_k \end{cases} \quad (4.1)$$

where $visit(\mathbf{b_k})$ is the number of times $\mathbf{b_k}$ has been visited and m_k is a parameter unique to each k. In other words, once a $\mathbf{b_k}$ is visited m_k times, we consider the estimate $M_k(\mathbf{b_k})$ reliable and assign $\hat{\pi}_k(\mathbf{b_k})$ to it. Henceforth we make no updates to $\hat{\pi}_k(\mathbf{b_k})$ (for $visit(\mathbf{b_k}) > m_k$). We discuss later (Equation 4.4) how m_k is chosen for each k. If a reliable estimate of $M_k(\mathbf{b_k})$ is unavailable (when $visit(\mathbf{b_k}) < m_k$), then $\hat{\pi}_k(\mathbf{b_k})$ is set to $\perp$ (meaning "I don't know").

Algorithm 3. MLES

---

input: $\epsilon, \delta, T, K_{max}$
;
repeat
   | Determine $\hat{\pi}_{best}$;;
   | Compute a policy using $\hat{\pi}_{best}$ (policy to follow for next T steps);;
   | $\tau \leftarrow 0$;
   | **repeat**
   | | Execute the policy;;
   | | $\tau \leftarrow \tau + 1$;;
   | **until** $\tau > T$;
   | Update all models based on the past T joint-actions;;
until *forever*;

MLES operates by planning for T steps at a time. The operations performed by MLES are as follows:

M1. Determine $\hat{\pi}_{best}$ (Line 3). Almost in every planning iteration assign the predictive model that best describes π_o as $\hat{\pi}_{best}$ by making a call to FIND-MODEL. However once in every $\lceil \frac{1-3\epsilon}{\epsilon} \rceil$ planning iterations, assign $\hat{\pi}_{best}$ by selecting randomly amongst the $K_{max} + 1$ models. The need of this exploratory iteration would become obvious once we specify our action selection mechanism in Section 4.1.4.

M2. Compute a stationary policy based on the $\hat{\pi}_{best}$ returned and execute it for the next T-steps (Lines 4 - 8). Note, a stationary policy in this case refers to a policy that is stationary in the context of an MDP, i.e., playing a fixed action for each state which may differ for different states.

M3. Update all models based on the past T joint-actions (Line 9).

Note the better the model returned in Step M1, the higher is the return accrued in Step M2. The main objective of Step M1 is then to consistently return a $\hat{\pi}_{best}$ which is a close approximation of π_o. That brings us to the concept of an ϵ-approx model for π_o.

Definition 19. *ϵ-approx model: We call a model $\hat{\pi}$ an ϵ-approx model of π_o, when for each feasible instantiation $\mathbf{b_K}$ of a bounded history of size K (i.e., $\mathbf{b_K} \in \mathcal{H}_K$), the prediction made by $\hat{\pi}$ is $\neq \perp$ and within a bound ϵ of $\pi_o(\mathbf{b_K})$.*

In order to have a close approximation of π_o, Step M1 relies on FIND-MODEL to return an $\frac{\epsilon}{T}$-approx model of π_o. An $\frac{\epsilon}{T}$-approx model of π_o is desired because the T-step expected return from following the optimal policy pertaining to such a model is always within ϵ of the T-step expected return from following the optimal policy pertaining to the true model π_o [45]. We next specify the details of FIND-MODEL, the main algorithmic component of MLES.

4.1.3 FIND-MODEL Algorithm

FIND-MODEL is the model selection algorithm running at the heart of MLeS. Its objective is to output the best predictive model for π_o from all possible $K_{max} + 1$ models maintained by MLeS.

Intuitively, all models of size $\geq K$ can learn π_o accurately (as they span over all of the past K joint-actions) with the bigger models requiring more samples to do so. On the other hand models of size $< K$ cannot fully represent π_o. From a high-level perspective, FIND-MODEL operates by comparing models of increasing size incrementally to determine the shortest most descriptive model such that all larger models cease to be more predictive of π_o. The next few paragraphs explain how FIND-MODEL functions. A reader not interested in deep technical details may directly skip to the paragraph before Lemma 4.1.1, our main theoretical result concerning FIND-MODEL. In short, Lemma 4.1.1 specifies the sufficient condition on exploration that needs to be satisfied for FIND-MODEL to return an $\frac{\epsilon}{T}$-approx model of π_o.

Since our approach involves comparing models of different sizes, we need some way of measuring how much they differ in their predictions. To that end we use a metric Δ_k.

Definition 20. Δ_k: Δ_k is the maximum difference in prediction between consecutive models of size k and $k + 1$. Let $Aug(\mathbf{b_k})$ be the set of all $k + 1$ length bounded histories which have $\mathbf{b_k}$ as the value of its first k joint-actions, and any possible value for the $k + 1$'st joint-action. Then,

$$\Delta_k = \max_{\mathbf{b_k}, \mathbf{b_{k+1}} \in Aug(\mathbf{b_k}))} ||\hat{\pi}_k(\mathbf{b_k}) - \hat{\pi}_{k+1}(\mathbf{b_{k+1}})||_\infty \ s.t. \ \hat{\pi}_{k+1}(\mathbf{b_{k+1}}) \neq \perp (4.2)$$

We will choose m_k's such that $\hat{\pi}_{k+1}(\mathbf{b_{k+1}}) \neq \perp$ will always imply $\hat{\pi}_k(\mathbf{b_k}) \neq \perp$. If for all $\mathbf{b_{k+1}}$'s, $\hat{\pi}_{k+1}(\mathbf{b_{k+1}}) = \perp$, then by default Δ_k is set to -1.

FIND-MODEL is fully specified in Algorithm 4. Its key steps are as follows.

S1. On every T-step planing iteration, for all $0 \leq k < K_{max}$, compute Δ_k (using Equation 4.2) and σ_k. If $\Delta_k = -1$, then we assign $\sigma_k = 1$. If $\Delta_k \neq -1$, then we assign σ_k as the tightest estimate satisfying the following condition:

$$\Pr(\Delta_k < \sigma_k) > 1 - \frac{\delta}{K_{max} + 1} \ \forall k \geq K \qquad (4.3)$$

By tightest we mean an estimate as close to Δ_k as possible. In such a case the σ_k is assigned as follows:

$$\sigma_k = \sqrt{\frac{1}{2m_k} log(\frac{2(K_{max} + 1)|A|N_k}{\delta})} + \sqrt{\frac{1}{2m_{k+1}} log(\frac{2(K_{max} + 1)|A|N_{k+1}}{\delta})}$$

Algorithm 4. FIND-MODEL

---

1   for all $0 \leq k < K_{max}$, compute Δ_k and σ_k;;
2   for $0 \leq k < K_{max}$ do
3       $flag \leftarrow$ true;;
4   for $k \leq k' < K_{max}$ do
5   if $\Delta_{k'} \geq \sigma_{k'}$ then
6             $flag \leftarrow$ false;;
7   break;;
8   if $flag$ then
9          $\hat{\pi}_{best} \leftarrow \hat{\pi}_k$;;
10   break;;

11   return $\hat{\pi}_{best}$;

N_k denotes the number of bounded histories of size k. The complete details on how we arrived at this is presented in Appendix A.1. Why we require the error probability from Equation 4.3 to be $\frac{\delta}{K_{max}+1}$ becomes apparent in the following step.

S2. FIND-MODEL then searches for that smallest value of k such that all the subsequent Δ_k's are less than their corresponding σ_k's (Lines 2 - 11 of Algorithm 4). It then concludes that this smallest k is the true value of K and returns $\hat{\pi}_k$ as $\hat{\pi}_{best}$. Since for each $k \geq K$, there is an error probability of at most $\frac{\delta}{K_{max}+1}$ with which the condition from Equation 4.3 may fail, the total error probability with which FIND-MODEL selects a model of size $\geq K$ remains upper-bounded by $\sum_{i=0}^{K_{max}} \frac{\delta}{K_{max}+1} = \delta$. Hence FIND-MODEL always selects a model of size at most K with a high probability of at least $1 - \delta$.

It is important to note that although we compute a σ_k for every $0 \leq k < K_{max}$, Equation 4.3 is only guaranteed to hold for $K \leq k < K_{max}$. However, in the early learning stages, Equation 4.3 may also hold for all $k \in [k', K_{max}\}$, where $k' < K$. This is generally true when we have not explored enough to deduce the relevance of all of the past K joint-actions. So initially FIND-MODEL may return sub-optimal models that are based on a shorter memory size than K. However once sufficient exploration has occurred (as quantified in the upcoming Lemma 4.1.1), then the model returned by FIND-MODEL is always an $\frac{\epsilon}{T}$-approx of π_o with high certainty.

We now state our main theoretical result concerning FIND-MODEL, namely Lemma 4.1.1. It states the sufficient condition on the exploration required to ensure that the $\hat{\pi}_{best}$ returned by FIND-MODEL is an $\frac{\epsilon}{T}$-approx of π_o, with a high likelihood. Complete details of all the calculations involved in the proof are presented in Appendix A.2. Recall that N_k denotes the number of bounded histories of size k, i.e., N_k is the size of $\mathcal{H}_k$.[3]

Lemma 4.1.1. *For any $0 < \epsilon < 1$ and $0 < \delta < 1$ and $m_K = O(\frac{K_{max}^2 T^2}{\epsilon^2}$ $log(\frac{K_{max} N_K |A|}{\delta}))$, once every $\mathbf{b_K} \in \mathcal{H}_K$ has been visited m_K times, the $\hat{\pi}_{best}$ returned by* FIND-MODEL *is of memory size at most K and an $\frac{\epsilon}{T}$-approx of π_o, with a high probability of at least $1 - 2\delta$.*

Thus it suffices to set m_k such that $\hat{\pi}_k$ stops predicting $\bot$ for a $\mathbf{b_k}$ as follows,

$$m_k = O(\frac{K_{max}^2 T^2}{\epsilon^2} log(\frac{K_{max} N_k |A|}{\delta})) \tag{4.4}$$

Lemma 4.1.1 gives us the condition that needs to be satisfied to ensure that the $\hat{\pi}_{best}$ returned by FIND-MODEL is an $\frac{\epsilon}{T}$-approx of π_o. However, it says nothing about how MLES should select its actions to ensure that this condition is satisfied. Next we focus on its action selection mechanism (Step M2) which ensures that the exploration condition from Lemma 4.1.1 holds.

4.1.4   Action Selection

In order to ensure that the condition of visits specified in Lemma 4.1.1 is met as quickly as possible, MLES uses an action selection mechanism based on the model-based RL algorithm RMAX [15]. We begin with a brief summary of how RMAX operates.

RMAX periodically computes a stationary policy by carefully balancing exploration and exploitation. The objective of the policy is to ensure faster exploration of state-action pairs that have not been visited many times, while ensuring a near optimal return if an accurate model of the MDP has already been learned. To encourage exploration of state-action pairs that have not been visited a sufficient number of times (say m), RMAX assigns an exploratory bonus to visiting that state-action pair. For state-action pairs that have been visited m times, RMAX performs the conventional Dynamic Programming backup. The policy is recomputed every time a new state-action pair is visited for the mth time.

There are two reasons why we choose RMAX as the RL algorithm for our action selection mechanism. First, its propensity to visit less visited states early in its learning stage is in line with our goal of achieving the necessary visits to all the $\mathbf{b_K}$'s (as recommended by Lemma 4.1.1) as early as possible. Second, it comes with a formal guarantee on the number of samples required

---

[3] We use N to denote the size of the state space of the underlying AIM. As our notion of AIM varies in different chapters depending on the other agents we are considering, so will our notion of N. With slight abuse of notation, we redefine N for each different context of an AIM.

to satisfy this exploration, which in turn facilitates our sample complexity bound for MLES.

MLES maintains a separate instance of RMAX for each of the possible $K_{max}+1$ AIMs corresponding to the $K_{max}+1$ possible models of π_o. At any iteration of MLES, let the $\hat{\pi}_{best}$ returned by Step M1 be $\hat{\pi}_k$ and the AIM associated with it be $\mathcal{M}_k$. MLES then picks the stationary policy computed from the RMAX instance associated with $\mathcal{M}_k$ to decide on the next T-step actions. The policy for the RMAX instance can be computed using any of the standard techniques, namely Value Iteration and Policy Iteration. MLES believes that k is the true value of K and hence attempts to explore all $\mathbf{b_k} \in \mathcal{H}_k$ m_k times to satisfy the condition of visits from Lemma 4.1.1. The policy computed from the RMAX instance associated with $\mathcal{M}_k$ precisely helps it to achieve that. However, there is a possibility that MLES might get stuck at a part of the state space where only some amongst the past K joint-actions are truly active (it may not reach up to K). In that case, it might converge to exploiting based on a sub-optimal model $\hat{\pi}_k$ and the return may then be far below U^*.

In order to avoid that, once in every $\lceil \frac{1-3\epsilon}{\epsilon} \rceil$ such T-step planning iterations, MLES computes the policy slightly differently. First, it chooses a k randomly from 0 to K_{max}. The goal in this iteration is to visit a new bounded history of size k which has not been visited m_k times. If such a visit is not possible (maybe because all such bounded histories have already been visited m_k times or they are reachable from the current start state with a very low probability), then exploit based on the current $\hat{\pi}_{best}$. The RMAX policy computation is then as follows. Assume that the state space of the underlying MDP comprised of all past K_{max} joint-actions. First, for all states of the MDP whose past k joint-actions have not been visited m_k times, provide them the exploratory bonus. For every other state use $\hat{\pi}_{best}$ to perform the Bellman back up. Note $\hat{\pi}_{best}$ only concerns itself with the joint-actions that are within its memory size and not on all of the past K_{max} joint-actions. Henceforth for future references, we call such a planning iteration as an *exploratory* iteration while the former a *greedy* iteration.

Now due to these exploratory iterations, $\hat{\pi}_K$ is chosen periodically as the random model in these exploratory iterations. Eventually by the implicit explore or exploit property of RMAX, it can be shown that at some exploratory iteration where MLES chooses $\hat{\pi}_K$ as the random model, it must achieve an expected return as high as $U^* - 2\epsilon$, with a high probability (since there are only finitely many entries to explore). Then from Assumption 1, we know that MLES must be following the optimal policy, otherwise such a high return would not have been possible. Thus MLES has learned a decent enough model of π_o that yields the optimal policy. Henceforth in every greedy iteration, it keeps exploiting based on this model and follows the optimal policy which eventually leads to a near optimal return. Complete details of how the above happens is presented in Appendix A.3 as the proof of the upcoming Lemma, our main theoretical result concerning MLES.

Lemma 4.1.2. *For any $0 < \epsilon < 1$ and $0 < \delta < 1$, with a high probability of at least $1 - 4\delta$, MLES achieves an actual return $\geq U^* - 5\epsilon$ against any memory-bounded o with memory size K, in a number of time steps given by*

$$O(\frac{N_K K_{max}^3 T^3}{\epsilon^7} log(\frac{K_{max} N_K |A|}{\delta}) log^2(\frac{1}{\delta})),$$

a quantity polynomial in $\frac{1}{\epsilon}$, $\frac{1}{\delta}$, K_{max}, N_K, $|A|$ and T.

Note our sample complexity bound is a worst case bound. Against most practical agents, MLES will likely converge to near optimal behavior in far fewer samples (as shown in our results section, Section 4.3).

The computational complexity of MLES for every planning iteration comprises two parts. The first part arises from FIND-MODEL, while the latter from the action selection step. FIND-MODEL takes an order of $O(K_{max}^2)$ computations on each planning iteration. For the action selection step, we need to solve a MDP. Hence the computational complexity for this step is equivalent to that of any MDP solver, such as Value Iteration.

This concludes our discussion on how MLES achieves targeted optimality against memory-bounded agents when Assumption 1 is satisfied. Next we show how Assumption 1 can be removed.

4.1.5 Removing Assumption 1

Our methodology follows the same line of reasoning as used by RMAX when it attempts to achieve a near optimal return in a MDP in polynomial sample complexity in cases where it is unaware of its desired planning horizon T. We first discuss how RMAX does so.

Let $P(T')$ denote the number of samples required by RMAX to achieve a near optimal return when the value of the planning horizon is T'. $P(T')$ is polynomial in T', the size of the state and action space of the MDP (namely the size of the MDP), as well as other relevant parameters. Being unaware of T, RMAX then repeatedly runs itself in restarts with incremental values of T, i.e., it first runs with $T = 1$, then with $T = 2$, and so forth. Whenever $P(T')$ time steps have elapsed since it started running with a planning horizon T', it stops and restarts with $T' + 1$. So eventually at some restart T' equals T and from that run onwards it always accrues a near optimal return. Since $\sum_{T'=1}^{T} P(T')$ is still polynomial in T, the size of the MDP and other relevant parameters, this technique of running RMAX in restarts still preserves its desired polynomial sample complexity property.

We use a similar technique when we lack a prior knowledge of a desired (ϵ, T) pair that satisfies Assumption 1. However there are a couple of subtle distinctions worth noting. First, unlike the case of RMAX we are unsure of the state space of the underlying AIM since we are unsure of the memory size. Second, we are dealing with two unknown values, namely ϵ and T, as opposed to just one for RMAX. Next we explain the modified MLES that addresses both of these problems with an emphasis on how it differs from the above presented version of RMAX.

The modified MLES algorithm operates as follows:

- We keep running Algorithm 3 in restarts with incremental values of T and decremental values of ϵ and δ. Let the values for T, ϵ and δ on run i be T_i, ϵ_i and δ_i respectively. We restart whenever Algorithm 3 has converged to a model and the number of time steps elapsed since it has converged to that model is equal to the sample complexity bound provided in Lemma 4.1.2. Note the latter requires a value of K which we get from our converged model. In each run i, Algorithm 3 always converges to a model that is at most of size K with an error probability of at most δ_i (from Lemma 4.1.1).
- Let T_i, δ_i and ϵ_i be assigned on the i'th run as follows:

$$T_i = 2^i \; , \; \delta_i = \frac{\delta_{init}}{2^i} \text{ and } \epsilon_i = \frac{\epsilon_{init}}{2^i}$$

where δ_{init} and ϵ_{init} are small initial probability values. Thus the total probability of ever selecting a model of size $> K$ is upper-bounded by $\sum_1^\infty \delta_i = \sum_1^\infty \frac{\delta_{init}}{2^i} = \delta_{init}$. So we have assured that our modified version of MLES (running Algorithm 3 in restarts) never ever operates on an AIM that is of memory size $> K$, with a high probability of at least $1 - \delta_{init}$.

- Furthermore, the number of runs required to reach the desired (ϵ, T) pair is upper-bounded by $max(\lceil log_2(T) \rceil, \lceil log_2(\frac{1}{\epsilon}) \rceil) + 1$. Suppose we reach our desired T earlier than our desired ϵ. Then the values of δ_i and T_i at the run when we reach our desired ϵ are,

$$\delta_i = \frac{\delta_{init}}{2^{\lceil log_2(\frac{1}{\epsilon}) \rceil + 1}} \approx O(\epsilon \delta_{init}) \text{ and } T_i = 2^{\lceil log_2(\frac{1}{\epsilon}) \rceil + 1} = O(\frac{1}{\epsilon})$$

On the contrary if we reach our desired ϵ earlier than our desired T, then the values of δ_i and ϵ_i at the run when we reach our desired T are,

$$\delta_i = \frac{\delta_{init}}{2^{\lceil log_2(T) \rceil + 1}} \approx O(\frac{\delta_{init}}{T}) \text{ and } \epsilon_i = \frac{\epsilon_{init}}{2^{\lceil log_2(T) \rceil + 1}} \approx O(\frac{\epsilon_{init}}{T})$$

Thus for each run until we reach the desired value of (ϵ, T) the sample complexity is polynomially dependent on the quantities listed in Lemma 4.1.2. Hence the total number of time steps elapsed until our modified version of MLES starts accruing a near optimal return is also polynomially dependent on the same quantities.

Now all that remains to be shown is how this modified version of MLES can be further improved to achieve safety. This can be achieved as follows. We always require that MLES (the modified version) checks its actual return before every restart. If the actual return is below $SV_i - \epsilon$, it plays its safety policy a sufficient number of time steps following it to compensate for the loss and bring the return back to within ϵ of SV_i, with a high probability of $1 - \delta$. The number of time steps it requires to play its safety policy to compensate for this loss is polynomial in the number of time steps for which that run lasted, $\frac{1}{\epsilon}$ and $\frac{1}{\delta}$. Hence before every restart, MLeS always achieves an actual return $\geq SV_i - \epsilon$ with a high probability of $1 - \delta$.

However by the definition of safety from [32], we require MLeS to ensure that there exists a $T > 0$ such that the expected return from any $T' \geq T$ steps of learning is provably within a desired bound of SV_i. What we show over here is that only at the beginning of any restart, MLeS achieves an actual return $\geq SV_i - \epsilon$ with a high certainty. What if the actual return falls below $SV_i - \epsilon$ in every run following a restart? Then we have not achieved safety. In this regard it can shown that after a certain number of restarts this never happens. In other words once we have ensured that the actual return remains $\geq SV_i - \epsilon$ for a certain number of restarts, then we have compensated enough to ensure that even if the learner achieves an actual return of zero in the next run, the overall actual return never falls below $SV_i - 2\epsilon$. Hence there exists a T such that MLeS always achieves an actual return $\geq SV_i - 2\epsilon$ with a high certainty, for any $T' > T$ time steps of learning. Hence safety is achieved by this modified version of MLeS. We point the reader to Appendix A.4 for a complete account of the details behind this claim.

This completes our complete analysis of MLeS. Next we present our full blown CMLeS algorithm.

4.2 Convergence with Model Learning and Safety (CMLeS)

CMLeS builds upon MLeS and achieves convergence as well, i.e., converges to a NE joint-policy in self-play. CMLeS begins by testing the other agents to see if they are also running CMLeS (self-play); when not, it uses MLeS as a subroutine. The algorithmic structure of CMLeS (Algorithm 5) comprises the following steps.

Lines 1 - 2: We assume that all agents have access to a NE solver and they compute a NE joint-policy. If the game has multiple NE joint-policies, CMLeS chooses randomly amongst them. So different CMLeS agents may settle for a different NE joint-policy;

Lines 3 - 4: CMLeS maintains a null hypothesis that all agents are following the same NE joint-policy ($AAPE$ = true). $AAPE$ stands for "all agents playing equilibrium". The hypothesis is not rejected unless CMLeS has evidence to the contrary. τ keeps count of the number of times the execution reaches Line 4;

Lines 5 - 8 : Whenever the algorithm reaches Line 5, it plays the equilibrium policy for a fixed number of time steps, N_τ. It keeps a running estimate of the empirical distribution of actions played by all agents, including itself, during this run. At Line 8, if for any agent j (including itself), the empirical distribution ϕ_j^τ differs from π_j^* by at least ϵ, $AAPE$ is set to false. The CMLeS agent has reason to believe that j may not be following the same

NE joint-policy that it computed. How the N_τ value is computed for each τ is explained in Section 4.2;

Lines 10 - 16: If $AAPE$ remains true after the execution of Line 8, the CM-LES agents continue to the next NE coordination phase by switching the execution back to Line 5. Once $AAPE$ is set to false, CMLES goes through a series of steps in which it checks whether the other agents are memory-bounded with memory size at most K_{max}. The details are explained below in Theorem 4.2.2. For the time being it suffices to know that the CMLES agents follow a fixed set of actions to signal to one another that they are indeed CMLES agents and in the process also detect K_{max} memory-bounded agents;

Lines 17 - 21: If all the agents follow the same fixed set of actions as described in Lines 10 -16, then CMLES sets $AAPE$ back to true and goes into a new NE coordination phase. For that it again computes a new NE joint-policy by choosing randomly from amongst the possible set of NE joint-policies;

Lines 22 - 23: Before the CMLES agents enter a new NE coordination phase, they check for each agent whether its actual return is below ϵ of its security value. If so, then that agent plays its safety policy for a sufficient number of time steps to compensate and ensure an actual return within ϵ of its security value, with a high probability of $1 - \delta$. Akin to MLES, the number of time steps it requires to play its safety policy depends polynomially on N_τ, $\frac{1}{\epsilon}$ and $\frac{1}{\delta}$. To keep every CMLES agent in sync, once a CMLES agent switches to following its safety policy to compensate for any loss, every other agent also does so, and waits for the process to complete. Once that is over, they go back and start a new NE coordination phase (Line 4);

Line 25: When the algorithm reaches here, it is sure (with probability 1) that the not all agents are following CMLES. Then it switches to following MLES;

Next we highlight some of CMLES's key theoretical properties.

Theoretical Underpinnings

We first show how N_τ is computed for each τ. Theoretically we want a N_τ such that if any agent j is following its share of a NE joint-policy π_j^*, then the empirical distribution of its actions over that N_τ time period (ϕ_j^τ) is always within ϵ of π_j^* with an error probability of at most $\frac{\delta}{2^{\tau+1}}$. We can compute that easily from Hoeffding bound,

$$N_\tau = O(\frac{|A|^2}{\epsilon^2} log(\frac{2^\tau}{\delta}))$$

Now we prove CMLES's first theoretical property.

Theorem 4.2.1. *In self-play, the* CMLES *agents converge to following a Nash equilibrium joint-policy in the limit.*

Algorithm 5. CMLES

---

input: $\epsilon, \delta, \tau = 0, \tau' = 0$

;

for $\forall j$ *in the set of agents* **do**
    $\pi_j^* \leftarrow$ ComputeNashEquilibriumStrategy()

$AAPE \leftarrow true$;

while $AAPE$ **do**
    for N_τ *time steps* **do**
       Play π_{self}^* ;
       For each agent j update ϕ_j^τ;

    recompute $AAPE$ using the ϕ_j^τ's and π_j^*'s;
    if $AAPE$ *is false* **then**
       if $\tau' = 0$ **then**
          Play a_i, $K_{max}+1$ times;

       else if $\tau' = 1$ **then**
          Play a_i, K_{max} times followed by a ;
          random action other than a_i;

       else
          Play a_i, $K_{max}+1$ times;

       if *all agents play the above prescribed set of actions* **then**
          $AAPE \leftarrow true$;
          $\tau' \leftarrow \tau' + 1$;
          for $\forall j$ *in the set of agents* **do**
             $\pi_j^* \leftarrow$ ComputeNashEquilibriumStrategy()

       if *actual return* $< SV_{self} - \epsilon$ **then**
          play safety policy enough times to compensate ;

    $\tau \leftarrow \tau + 1$

Play MLES

Proof. The proof follows in three parts. The first part of the proof shows that in self-play the execution of CMLES always reaches Line 5 once it reaches Line 10. That is the CMLES agents get infinite number of chances to coordinate to a NE joint-policy. This holds because once its execution reaches Line 10, it follows a fixed set of prescribed actions. Since each CMLES agent follows this fixed policy, CMLES remains assured that all other agents are indeed CMLES agents and its execution reaches Line 5 to start a new NE coordination phase.

The second part shows that all the CMLES agents compute the same NE joint-policy periodically. This is very easy to show. If there are k different NE joint-policies and n agents, then in expectation once in every k^n NE coordination phases, the CMLES agents must choose the same NE joint-policy.

The third part shows that in self-play, the probability of all the CMLES agents following a NE joint-policy forever, once they select the same one in some NE coordination phase, is non-zero and this probability increases monotonically with every such NE coordination phase (where they select the same NE joint-policy). This is ensured by our choice of N_τ for each τ. Assume that the first time when they compute the same NE joint-policy is for a NE coordination phase with $\tau = p$. From union bound, it can be shown that the probability of $AAPE$ ever getting set to false from that NE coordination phase and onwards is upper bounded by $n \sum\limits_{\tau=p}^{\infty} \dfrac{\delta}{2^{\tau+1}} = \dfrac{n\delta}{2^p}$. Let the next NE coordination phase when all agents compute the same NE joint-policy be q ($q > p$). The probability of $AAPE$ ever getting set to false from that NE coordination phase and onwards is upper bounded by $n \sum\limits_{\tau=q}^{\infty} \dfrac{\delta}{2^{\tau+1}} = \dfrac{n\delta}{2^q}$, and so forth. Since $\frac{n\delta}{2^p} > \frac{n\delta}{2^q}$, the claim follows.

Combining these three parts of the proof, it follows that in infinite repeated play, CMLES converges to following a Nash equilibrium joint-policy in the limit. □

CMLES cannot distinguish between a CMLES agent and a memory-bounded agent if the latter by chance plays the computed NE joint-policy from the beginning, and may coordinate with it to converge to a NE. Note, this might not strictly be the best response against such a group of agents, but we believe it is still a reasonable solution concept. Henceforth, our analysis on memory-bounded agents will exclude this special case.

Theorem 4.2.2. CMLES *achieves targeted optimality against memory-bounded agents whose memory size is upper-bounded by a known value K_{max}.*

Proof. The proof follows in two parts. In the first part, we argue that given these other agents are not following the NE joint-policy, every time the execution reaches Line 5, there is a non-zero probability that it reaches Line 10. This part of the proof follows trivially from how we compute $AAPE$ in Line 8.

The second part of the proof shows that given the execution reaches Line 10 periodically, the execution must eventually reach Line 25 at some point and switch to following MLES. We utilize the property that a K memory-bounded agent is also a K_{max} memory-bounded agent. The first time $AAPE$ is set to false, CMLES selects a random action a_i and then plays it $K_{max}+1$ times in a row. The second time when $AAPE$ is set to false, it plays a_i, K_{max} times followed by a different action. If the other agents have behaved identically in both of the above situations, then CMLES knows : 1) either the rest of the agents are following CMLES, or, 2) if they are memory-bounded with a memory size upper-bounded by K_{max}, they play stochastically for a K_{max} bounded memory where all agents play a_i. The latter observation comes in handy below. Henceforth, whenever $AAPE$ is set to false, CMLES always plays a_i, $K_{max}+1$ times in a row. Since a memory-bounded agent must be stochastic (from the above observation), at some point of time, it will play a different action on the $K_{max}+1$'th step with a non-zero probability. CMLeS then rejects the null hypothesis that all other agents are CMLeS agents and jumps to Line 25.

Combining these two parts of the proof, it follows that against memory-bounded agents whose memory size is upper-bounded by K_{max}, CMLeS eventually converges to following MLeS and hence achieves targeted optimality.   □

All that remains to be shown is that CMLeS achieves safety against arbitrary agents. If CMLeS converges to following MLeS, then by virtue of MLeS, it achieves safety. If CMLeS never converges to following MLeS, then Lines 22 - 23 ensure that at the beginning of any NE coordination phase, it always achieves an actual return $\geq SV_i - \epsilon$ with a high probability of $1 - \delta$. It can shown that after a certain number of NE coordination phases, we compensate enough to ensure that even if CMLeS achieves an actual return of zero in the next coordination phase, the overall actual return never falls below $SV_i - 2\epsilon$ (analogous to the proof in Appendix A.4). Hence safety is achieved by CMLeS.

4.3   Results

Whereas the main contribution of this chapter is the introduction of CMLeS as a theoretically grounded MAL algorithm, we would also like it to be useful in practice. As an empirical exercise, we choose to focus on how efficiently MLeS (the main component of CMLeS) models memory-bounded agents in comparison to existing algorithms, Pcm(A) and Awesome. Our empirical analysis uses the version of MLeS presented in Algorithm 3, not the one which runs in restarts.

Theoretically, the specification of MLeS depends on the following input parameters: δ, ϵ, T and K_{max}. δ, ϵ and T together determine the m_k and σ_k for each model. Recall that m_k is the number of visits we require to each $\mathbf{b_k}$ to consider the estimate $M_k(\mathbf{b_k})$ (empirical distribution of o's play for a k sized bounded history $\mathbf{b_k}$) reliable. Furthermore we require the (ϵ, T) pair to satisfy Assumption 1. An implementation of MLeS straight from its theoretical specification is challenging for the following reasons.

1. First, there exists no principled way of guessing an (ϵ, T) pair a priori that satisfies Assumption 1.
2. Second, even if we know such an (ϵ, T) pair, the value of each m_k computed based on it is prohibitively high for practical purposes. Note, the definition of m_k is a very conservative one and in practice much smaller values of m_k should suffice.

Hence we introduce a few approximations when implementing MLeS. First, instead of seeding MLeS with a δ, ϵ, K_{max} and T, we seed it with an m, δ and K_{max}. m plays the role of m_k and is the same for models of all sizes. δ is required to compute the value of σ_k (Equation A.3). All our results are reported for $m = 20$ and $\delta = 0.2$. K_{max} in our case consists of the past 10 joint-actions. In other words, we let MLES figure out which amongst these past 10 joint-actions can be best used to model the other agents.

Also, note that MLES needs an exploratory iteration once every $\lceil \frac{1-3\epsilon}{\epsilon} \rceil$ planning iterations. Since we do not specify a value for ϵ, it is not clear when to opt for an exploratory iteration. Hence we opt against an exploratory iteration. In all of our experiments, the explorations that happen in the greedy iterations are sufficient to generate good results.

Finally, MLES functions by planning for T time steps at a time (see Algorithm 3). Such a T-step action selection policy is just the stationary RMAX policy computed by running Value Iteration in the underlying AIM and executed for T steps. In our implementation, we keep executing the computed stationary RMAX policy forever, unless a new state of the underlying AIM gets visited for the m'th time. In that case, we recompute it. This approach is structurally similar to the one described in Algorithm 3, except that it is more computationally efficient.

We use the 3-player Prisoner's Dilemma (PD) game as our representative matrix game. The game is a 3 player version of the n-player PD present in gamut.[4] In this version, the payoff to each agent is based on the number of agents who "cooperate" not including the agent itself. If the number of other agents who "cooperate" is i, then we say that $C(i)$ is the payoff for cooperating and $D(i)$ is the payoff for defecting. In order for this payoff scheme to result in a Prisoners Dilemma, it must be the case that:

- $D(i) > C(i)$ for $0 \leq i \leq n - 1$;
- $D(i + 1) > D(i)$ and also $C(i + 1) > C(i)$ for $0 \leq i < n - 1$;
- $C(i) > \frac{D(i)+C(i-1)}{2}$ for $0 < i \leq n - 1$;

The payoffs supporting this payoff scheme is automatically generated by gamut and so we need not worry about it. The memory-bounded strategies we test against are,

Type 1: every other player plays "defect" if in the last 5 steps MLES played "defect" even once. Otherwise, they play " cooperate". The agents are thus deterministic memory-bounded strategies with $K = 5$;

Type 2: every other player behaves as type-1 with 0.5 probability, or else plays completely randomly. In this case, the agents are stochastic with $K = 5$;

The total number of bounded histories of size 10 in this case is 8^{10}, which makes PCM(A) highly inefficient. However, MLES quickly figures out the true K and converges to the optimal behavior in a reasonable number of steps. Figure 4.3 shows our results against these two types of agents. The

---

[4] http://gamut.stanford.edu

Y-axis shows the payoff of each algorithm as a fraction of the optimal payoff achievable against the respective opponent. Each plot has been averaged over 30 runs to increase robustness. Against type-1 agents (Figure 4.3(i)), MLeS figures out the true memory size in about 2000 steps and converges to playing near optimally by 20000 episodes. Against type-2 agents (Figure 4.3(ii)), it takes a little longer to converge to playing near optimally (about 30000 episodes) because in this case, the number of feasible bounded histories of size 5 are much more. Both AWESOME and PCM(A) perform much worse.

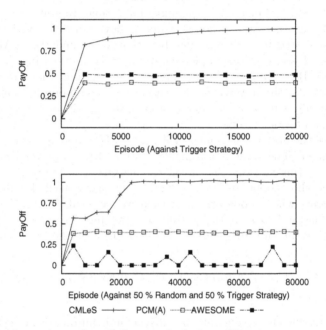

Fig. 4.1 Against memory-bounded agents in 3 player PD.

Against type-1 agents (upper figure), MLeS figures out the true memory size in about 2000 steps and converges to playing near optimally by 20000 episodes. Against type-2 agents (bottom figure) MLeS converges to playing near optimally in about 30000 episodes. Both AWESOME and PCM(A) perform much worse.

4.4 Summary

In this chapter, we introduced a novel MAL algorithm, CMLeS, which in an arbitrary repeated game, achieves convergence, targeted-optimality against memory-bounded agents whose memory size is upper-bounded by a known value K_{max}, and safety. A key contribution of CMLeS is in the manner it handles memory-bounded agents: it requires only a loose upper bound of the other agents' memory size. Second, and more importantly, CMLeS improves upon the state of the art algorithm, by promising targeted optimality against memory-bounded agents by requiring sufficient number of visits to only all feasible joint histories of size K, where K is the other agents' true memory size. This concludes our formal treatment of memory-bounded agents in this book. Starting with the following chapter, we concern ourself with a significantly more complex class of agents, namely Markovian agents.

Chapter 5
Maximizing Social Welfare in the Presence of Markovian Agents

In the previous chapter, we proposed a MAL algorithm CMLeS that in a arbitrary repeated game, converges to following a NE joint-policy in self-play, achieves close to the best response with a high probability against a set of memory-bounded agents whose memory size is upper-bounded by a known value, and achieves close to the security value against any other set of agents which cannot be represented as being K_{max} memory-bounded. CMLeS is the first MAL algorithm to achieve all of the above objectives.

However there are two shortcomings of CMLeS. First, it requires all the CMLeS agents to compute a NE to ensure that they converge to one in self-play. Computing a NE for arbitrary matrix games is a computationally hard problem. In fact it is known to be PPAD complete [24]. Second, the only guarantees of optimality it provides is against a homogeneous set of memory-bounded agents whose memory size is upper-bounded by a known value K_{max}. In a population comprised of more than one CMLeS agent and the rest memory-bounded agents with a memory size bound of K_{max}, from a single CMLeS agent's perspective, the rest of the population is still arbitrary (as it does not comprise wholly of CMLeS or memory-bounded agents). In such a scenario, CMLeS only ensures a return close to the security value. As one might expect if the CMLeS agents can coordinate with each other, they can guarantee a better joint-return.

The main contribution of this chapter (the second big contribution of this book) is to introduce a novel MAL algorithm called JOMA (*Joint Optimization against Markovian Agents*) that does not need to solve a computationally hard problem to achieve its objective and can also coordinate with other self agents to jointly explore and exploit. Our initial analysis caters to a simpler setting than that considered in Chapter 4: we assume that the JOMA agents know each others' identities in the population. We show that in such a setting, JOMA achieves the following two objectives:

D. Chakraborty, *Sample Efficient Multiagent Learning in the Presence* 49
of Markovian Agents, Studies in Computational Intelligence 523,
DOI: 10.1007/978-3-319-02606-0_5, © Springer International Publishing Switzerland 2014

- achieves close to the social welfare maximizing joint-return in a population comprised of self (JOMA) and Markovian agents, in provably efficient sample complexity. The social welfare value is the sum of returns of the JOMA agents in the population, the agents we control;[1]
- ensures safety against any other set of agents;

The complete technical details behind how JOMA achieves these two objectives is presented in Sections 5.2 to 5.4.

In the second part of our analysis, we show how JOMA fairs in more general settings where it does not a-priori know the identities of the other JOMA agents in the population. In such settings, we provide similar guarantees as the ones cited above except for a minor, yet necessary relaxation. We address this in detail in Section 5.5.

As introduced in Definition 16, a Markovian agent chooses its actions as a (fixed) function of a set of discrete feature variables computed from the joint history of play. Depending on the joint-action taken, these feature values transition in a Markovian fashion on every time step (see Definition 16 from Chapter 2). We assume that JOMA has some prior knowledge of the possible set of features upon which the Markovian agents may base their policies, but not the exact set. In order to achieve its first objective, each JOMA agent coordinates with the other JOMA agents in the population to figure out the relevant features from this feature set that determine the Markovian agents' policies and then exploits them to maximize social welfare. JOMA does not need to solve any computationally hard problem to achieve this goal. Also recognizing that each agent needs to be individually motivated to follow JOMA, we respect the requirement that each JOMA agent achieve at least its security value. Recall that each agent can always play its safety policy to unilaterally ensure its security value. Thus in an attempt to maximize social welfare, if some JOMA agent receives an individual return less than its security value, then the other JOMA agents (agents with an individual return greater than their security values) compensate by making *side payments* [16] to that agent. The mechanism of side payment in place is the standard one as defined in [16] and hence we omit further details.

JOMA represents a significant step forward in the theory of MAL by introducing a novel mechanism of modeling a comparatively more complex class of agent behavior than that has been modeled to date. To the best of our knowledge, it is the first MAL algorithm that models other agents based on features. In many real world problems, we have prior knowledge of decent features that may characterize the other agents' policies. JOMA can utilize such prior knowledge to good effect in modeling such agents. However, often such features are non-Markovian and then we cannot frame the learning problem as learning in an unknown MDP. Even in such scenarios, leveraging

---

[1] Our result naturally extends to the general notion of social welfare as defined in Definition 11. However in this setting our current definition of SW seems more natural.

from existing planning techniques that perform in non-Markovian settings, JOMA can be applied to solve such problems. We give a couple of examples of such scenarios in the next chapter.

Though in this chapter we focus primarily on characterizing and proving JOMA's theoretical properties, we also provide some initial empirical results in which it compares favorably to some of its peers from literature. The remainder of the chapter is organized as follows. Section 5.1 provides a road map on how we incrementally introduce JOMA, Sections 5.2 to 5.5 present all the algorithmic aspects of JOMA, Section 5.6 presents some empirical validations of the theoretical properties of JOMA and Section 5.7 summarizes the chapter.

5.1   Road Map to Specifying JOMA

The incremental steps that lead to fully specifying JOMA are as follows. In the first three steps, we assume that the JOMA agents know each other and only need to model the Markovian agents in the population. In the fourth step, we remove this assumption. We also assume that JOMA has some prior knowledge of the possible set of features, denoted by $\mathcal{F}$, upon which the Markovian agents may base their policies, but not the exact set $F \subseteq \mathcal{F}$.

Step 1. The first step introduces a subroutine that solves a much simpler, yet challenging problem (Section 5.2). Here we make a couple of simplifying assumptions,

1. There is only one JOMA agent in the population, the rest being Markovian. We denote the set of Markovian agents as a single agent o with an unknown policy π_o. Note, a set of Markovian agents can be treated as a single Markovian agent whose policy and action space is just the joint-policy and joint-action space of all of them respectively. For the sake of simplicity in analysis, we assume that the action space of o is A;

2. The feature set $\mathcal{F}$ that determines the unknown policy of the Markovian agents (π_o) satisfies a sequential structure, i.e., the features in $\mathcal{F}$ are arranged in a sequence such that the first K features completely determine π_o, K being unknown;[2]

The objective of the subroutine is to achieve an actual return very close to the optimal return (the return from optimally exploiting π_o), with a high likelihood.

Step 2. The second step builds on the first step and provides a subroutine that solves a progressively harder problem (Section 5.3). We again assume that there is only one JOMA agent in the population, the rest being

---

[2] Note, in this chapter K refers to the number of relevant features and should not be confused with the memory size of a memory-bounded agent as used in Chapters 3 and 4.

Markovian. However the second assumption from Step 1 no longer holds (the features in $\mathcal{F}$ can be arranged in any order). Again the objective of this subroutine is to achieve an actual return very close to the optimal return, with a high likelihood (given that the sequential structure assumption does not hold).

Step 3. In the third step we specify the full blown JOMA algorithm (Section 5.4). The population comprised of more than one JOMA agent, the rest being Markovian agents. There are two high-level algorithmic components of JOMA:

1. First, it goes through an ordering computation phase where it settles on a shared common ordering amongst all the JOMA agents, with all other JOMA agents;
2. Second, it reverts to using the subroutine from Step 2 to achieve an actual joint-return very close to the social welfare maximizing joint-return;

JOMA achieves both of the above subgoals with a very high likelihood.

Step 4. Finally in the fourth step, we analyze the case when the JOMA agents are unaware of each others' identities (Section 5.5). Our specification of JOMA remains the same as introduced in Step 3, however our analysis changes.

We next delve deep into specifying our algorithms for each of the above steps.

5.2 Model Learning and Exploitation of Markovian Agents with the Sequential Structure Assumption

In this section we present our subroutine for Step 1 of our road map. As before we denote the agent under our control as i and the Markovian agents as a single agent o, with π_o being their unknown policy. Let $\mathcal{F}$ be the set of possible features that π_o might depend on and n be its size.[3] Recall that in this case we assume that the feature set $\mathcal{F}$ satisfies the sequential structure assumption. Hence we call the associated learning problem the sequential structure learning problem.

Definition 21. *Sequential Structure Learning problem: In a sequential structure learning problem, the features in $\mathcal{F}$ are arranged in a sequence such that all of the K relevant features that determine π_o precede the irrelevant features in the sequence, with K being unknown. The learning problem is how to efficiently model π_o given this rich feature space representation and achieve a near optimal return with a high likelihood, in efficient sample complexity.*

---

[3] Note, in this chapter n denotes the number of features in $\mathcal{F}$ and should not be confused with the number of agents playing the repeated game as introduced in Definition 7. We do not use the latter in our analysis in this chapter.

A careful reader would observe that the problem of achieving targeted optimality against a memory-bounded agent whose memory size is upper-bounded by K_{max} is an instance of the sequential structure learning problem. The following explains why. Consider a memory-bounded agent o with memory size K. Recall from Chapter 2 that a memory-bounded agent is a special case of a Markovian agent. So the last K joint-actions completely determine π_o. Assume that i is unaware of the exact value of K, but is prepared to model o with a memory size $K_{max} \geq K$. Let the features in $\mathcal{F}$ be the most recent K_{max} joint-actions. As the first K features (K being unknown to i) from $\mathcal{F}$ completely determine π_o and all of the remaining features are irrelevant, the learning problem is a sequential structure learning problem. Thus our approach to solving the sequential structure learning problem is very similar to our approach of achieving targeted optimality against a memory-bounded agent whose memory size is upper-bounded by K_{max}. However for the sake of completeness, we present it in detail.

Though the sequential structure assumption seems restrictive, we later show how a solution to this simpler, yet nontrivial, learning problem results in an efficient solution to the general learning problem where such an ordering is unknown. We call our subroutine that solves the sequential structure learning problem, *M*odel *L*earning and *E*xploitation against *M*arkovian agents with the *S*equential structure assumption (MLeM(S)).

We can always model π_o by assuming that it is Markovian based on the entire set of features $\mathcal{F}$. Our new notion of a *model* $\hat{\pi}$ for π_o is defined as follows.

Definition 22. *Model: A model* $\hat{\pi} : F' \mapsto \Delta A$ *of* π_o *is defined over a set of parent features* $F' \subseteq \mathcal{F}$, *and specifies a distribution over the action set A (mixed action) for every feasible instantiation of the parent features.*

Modeling π_o based on the entire $\mathcal{F}$ may involve learning over a much larger state space than is necessary. Our goal is to model π_o with the shortest most descriptive model and hence achieve the lowest possible sample complexity bound by avoiding unnecessary exploration.

MLeM(S) is introduced in Algorithm 6. Just as in the case of MLeS (Section 4.1), for the sake of clarity we break our algorithmic analysis of MLeM(S) into four parts. First in Section 5.2.1, we discuss the choice of the inputs for MLeM(S). Second in Section 5.2.2, we describe how MLeM(S) operates from a high-level. Third and fourth in Sections 5.2.3 and 5.2.4 respectively, we focus on MLeM(S)'s two main algorithmic components: the FIND-MODEL(S) algorithm and its action selection mechanism.

5.2.1 Inputs to MLEM(S)

The inputs to MLeM(S) are ϵ, δ, T and $\mathcal{F}$. Both ϵ and δ are small probability values. T is the planning horizon explained in the next paragraph. A reader not interested in a deep theoretical understanding of MLeM(S) may skip the rest of this subsection and treat these inputs as free parameters. We devote the rest of this subsection to justifying the choice behind these input parameters that facilitates our theoretical claims concerning MLeM(S).

Just like MLeS, MLeM(S) operates by planning for T time steps at a time. In each such planning iteration, it uses the best model of π_o at hand and plans its actions for the next T time steps based on it. Let U^* be the expected return from the best response against o, i.e., the optimal return achievable in the AIM induced by π_o. To facilitate the theory behind our claim that MLeM(S) converges to follow the best response against o, we assume that the (ϵ, T) pair taken as input always satisfies Assumption 1 from Section 4.1.1. We restate it again for the reader's convenience.

The planning horizon T is sufficiently large and the ϵ sufficiently small to ensure that

1. *T is the ϵ-return mixing time of the optimal policy for the AIM;*
2. *for any sub-optimal policy π and for any state s of the induced AIM, $U_T^\pi(s) < U^* - 2\epsilon$;*

We refer the reader to Section 4.1.1 for a detailed discussion on the existence of such an (ϵ, T) pair. Our analysis in this chapter assumes that MLeM(S) (and also its successive variants introduced later in the chapter) is aware of such an (ϵ, T) pair that satisfies Assumption 1. In cases where it is unaware of such an (ϵ, T) pair, the approach presented in Section 4.1.5 which extends MLeS to solve this general case, applies here as well. Note, the above approach by default ensures safety. Having noted that we skip an analysis for the case when the (ϵ, T) pair is unknown (and also how MLeM(S) achieves safety).

5.2.2 High Level Idea of MLeM(S)

Now we move on to the high level idea of MLeM(S) (Algorithm 6) which is very similar to that of MLeS. Since MLeM(S) is unaware of the exact K that characterizes π_o, it maintains a model of π_o for each of the first k features from $\mathcal{F}$, $k \in [0 : n]$. Thus it maintains $n + 1$ models in total.

Let the model that is based on the first k features be $\hat{\pi}_k$. Internally each $\hat{\pi}_k$ maintains a value $M_k(\mathbf{b_k})$ which is the maximum likelihood distribution of o's play for every feasible value $\mathbf{b_k}$ of the first k features from $\mathcal{F}$.[4] Whenever the first k features assume a value $\mathbf{b_k}$ in online play, we say a *visit* to $\mathbf{b_k}$ has occurred. $\hat{\pi}_k(\mathbf{b_k})$ is then defined as follows:

$$\hat{\pi}_k(\mathbf{b_k}) = \begin{cases} M_k(\mathbf{b_k}) & \text{once } visit(\mathbf{b_k}) = m_k \\ \bot & \text{when } visit(\mathbf{b_k}) < m_k \end{cases} \tag{5.1}$$

---

[4] Note, in this chapter $\mathbf{b_k}$ denotes a value of the first k features from $\mathcal{F}$, not a value of the past k joint-actions as used in Chapter 4.

Algorithm 6. MLEM(S)

---

input: $\epsilon, \delta, T, \mathcal{F}$

;

repeat

    Determine $\hat{\pi}_{best}$;;

    Compute a policy using $\hat{\pi}_{best}$ (policy to follow for next T steps);;

    $\tau \leftarrow 0$;

    repeat

        Execute the policy;;

        $\tau \leftarrow \tau + 1$;;

    until $\tau > T$;

    Update all models based on the past T joint-actions;;

until *forever*;

---

where $visit(\mathbf{b_k})$ is the number of times $\mathbf{b_k}$ has been visited and m_k is a parameter unique to each k. In other words, once a $\mathbf{b_k}$ is visited m_k times, we consider the estimate $M_k(\mathbf{b_k})$ reliable and assign $\hat{\pi}_k(\mathbf{b_k})$ to it. Henceforth we make no updates to $\hat{\pi}_k(\mathbf{b_k})$ (for $visit(\mathbf{b_k}) > m_k$). We discuss later (Equation 5.5) how m_k is chosen for each k. If a reliable estimate of $M_k(\mathbf{b_k})$ is unavailable (when $visit(\mathbf{b_k}) < m_k$), then $\hat{\pi}_k(\mathbf{b_k})$ is set to $\bot$ (meaning "I don't know").

MLEM(S) operates by planning for T steps at a time. The operations performed by MLEM(S) are as follows:

M1. Determine $\hat{\pi}_{best}$ (Line 2). Almost in every planning iteration assign the predictive model that best describes π_o as $\hat{\pi}_{best}$ by making a call to FIND-MODEL(S). However once in every $\lceil \frac{1-3\epsilon}{\epsilon} \rceil$ planning iterations, assign $\hat{\pi}_{best}$ by selecting randomly amongst the $n+1$ models. The need of this exploratory iteration would become obvious once we specify our action selection mechanism in Section 5.2.4.

M2. Compute a stationary policy based on the $\hat{\pi}_{best}$ returned and execute it for the next T steps (Lines 3 - 8).

M3. Update all models based on the past T joint-actions (Line 9).

Note the better the model returned in Step M1, the higher is the return accrued in Step M2. The main objective of Step M1 is then to consistently return a $\hat{\pi}_{best}$ which is a close approximation of π_o. That brings us to our new concept of an ϵ-approx model for π_o (as opposed to the ϵ-approx concept from Chapter 4, Definition 19).

Definition 23. *ϵ-approx model: We call a model $\hat{\pi}$ an ϵ-approx model of π_o, when for each feasible value $\mathbf{b_K}$ of the relevant feature set $\{f_1, \ldots, f_K\}$, the following condition holds:*

$$\hat{\pi}(\mathbf{b_K}) \neq \bot \wedge ||\hat{\pi}(\mathbf{b_K}) - \pi_o(\mathbf{b_K})||_\infty \leq \epsilon \tag{5.2}$$

In order to have a close approximation of π_o, Step M1 relies on FIND-MODEL(S) to return an $\frac{\epsilon}{T}$-approx model of π_o. An $\frac{\epsilon}{T}$-approx model of π_o is desired because the T-step expected return from following the optimal policy pertaining to such a model is always within ϵ of the T-step expected return from following the optimal policy pertaining to the true model π_o [45]. We next specify the details of FIND-MODEL(S), the main algorithmic component of MLEM(S).

5.2.3 FIND-MODEL(S) Algorithm

FIND-MODEL(S) is the main algorithm running at the heart of MLEM(S) and is the key to understanding all of our theoretical claims in this section. Its objective is to output the best predictive model for π_o from all possible $n+1$ models maintained by MLEM(S). Again its functioning is very similar to the FIND-MODEL algorithm of MLES from Section 4.1.3.

Intuitively, all models of size $\geq K$ can learn π_o accurately (as they consist of all of the relevant features) with the bigger models requiring more samples to do so. On the other hand models of size $< K$ cannot fully represent π_o. From a high-level perspective, FIND-MODEL(S) operates by comparing models of increasing size incrementally to determine the shortest most descriptive model such that all larger models cease to be more predictive of π_o. The next few paragraphs explain how FIND-MODEL(S) functions. A reader not interested in deep technical details may directly skip to the paragraph before Lemma 5.2.1, our main theoretical result concerning FIND-MODEL(S). In short, Lemma 5.2.1 specifies the sufficient condition on exploration that needs to be satisfied for FIND-MODEL(S) to return an $\frac{\epsilon}{T}$-approx model of π_o.

Since our approach involves comparing models of different sizes, we need some way of measuring how much they differ in their predictions. To that end we use a new Δ_k metric (as opposed to the Δ_k metric from Chapter 4, Definition 20).

Definition 24. Δ_k: Δ_k is the maximum difference in prediction between consecutive models of size k and $k+1$. Let $Aug(\mathbf{b_k})$ be the set of all $k+1$ length vectors which have $\mathbf{b_k}$ as the value of their first k features, and a feasible value of the $k+1$'th feature from $\mathcal{F}$ as its $k+1$'st value. Then,

$$\Delta_k = \max_{\mathbf{b_k}, \mathbf{b_{k+1}} \in Aug(\mathbf{b_k}))} ||\hat{\pi}_k(\mathbf{b_k}) - \hat{\pi}_{k+1}(\mathbf{b_{k+1}})||_\infty \; s.t. \; \hat{\pi}_{k+1}(\mathbf{b_{k+1}}) \neq \perp \text{(5.3)}$$

We will choose m_k's such that $\hat{\pi}_{k+1}(\mathbf{b_{k+1}}) \neq \perp$ will always imply $\hat{\pi}_k(\mathbf{b_k}) \neq \perp$. If for all $\mathbf{b_{k+1}}$'s, $\hat{\pi}_{k+1}(\mathbf{b_{k+1}}) = \perp$, then by default Δ_k is set to -1.

FIND-MODEL(S) is fully specified in Algorithm 7 and is identical to Algorithm 4 from Chapter 4, except for the obvious differences in the manner the Δ_k's and the models are computed. Nonetheless for the sake of

---

Algorithm 7. FIND-MODEL(S)

$\hat{\pi}_{best} \leftarrow \hat{\pi}_n$;
1   for all $0 \le k < n$, compute Δ_k and σ_k;
2   for $0 \le k < n$ do
3 | $flag \leftarrow$ true;
4 | for $k \le k' < n$ do
5 | | if $\Delta_{k'} \ge \sigma_{k'}$ then
6 | | | $flag \leftarrow$ false;
7 | | break;
8 | if $flag$ then
9 | | $\hat{\pi}_{best} \leftarrow \hat{\pi}_k$;
10 | break;

11 return $\hat{\pi}_{best}$

---

completeness, we present it in details in the following paragraphs. Its key steps are as follows.

S1. On every T step planning iteration, for all $0 \le k < n$, compute Δ_k (using Equation 5.3) and σ_k. If $\Delta_k = -1$, then we assign $\sigma_k = 1$. If $\Delta_k \neq -1$, then we assign σ_k as the tightest estimate satisfying the following condition:

$$\Pr(\Delta_k < \sigma_k) > 1 - \frac{\delta}{n+1} \ \forall k \ge K \qquad (5.4)$$

By tightest we mean an estimate as close to Δ_k as possible. The details on how the σ_k's are computed are identical to the case of MLES and are presented in Appendix A.1. Why we require the error probability from Equation 5.4 to be $\frac{\delta}{n+1}$ becomes apparent in the following step.

S2. FIND-MODEL(S) then searches for that smallest value of k such that all the subsequent Δ_k's are less than their corresponding σ_k's (Lines 2 - 11). It then concludes that this smallest k is the true value of K and returns $\hat{\pi}_k$ as $\hat{\pi}_{best}$. Since for each $k \ge K$, there is an error probability of at most $\frac{\delta}{n+1}$ with which the condition from Equation 5.4 may fail, the total error probability with which FIND-MODEL(S) selects a model of size $\ge K$ remains upper-bounded by $\sum_{i=0}^{n} \frac{\delta}{n+1} = \delta$. Hence FIND-MODEL(S) always selects a model of size at most K with a high probability of at least $1 - \delta$.

It is important to note that although we compute a σ_k for every $0 \le k < n$, Equation 5.4 is *guaranteed* to hold only for $K \le k < n$. However as observed earlier, in the early learning stages Equation 5.4 may also hold for all $k \in [k', n\}$, where $k' < K$. This is generally true when we have not explored enough to deduce the relevance of all K features. So initially FIND-MODEL(S) may return sub-optimal models. However once sufficient exploration has occurred (as quantified in the upcoming Lemma 5.2.1), the model returned by FIND-MODEL(S) will be an $\frac{\epsilon}{T}$-approx of π_o with high certainty.

We now state our main theoretical result concerning FIND-MODEL(S), namely Lemma 5.2.1. It states the sufficient condition on the exploration required to ensure that the $\hat{\pi}_{best}$ returned by FIND-MODEL(S) is an $\frac{\epsilon}{T}$-approx of π_o, with a high likelihood. Complete details of the proof behind the Lemma are identical to Lemma 4.1.1 of MLES (except how the Δ_k's and the models are computed) and are presented in Appendix A.2. Henceforth N_k (redefined) denotes the number of feasible values of the feature set $\{f_1, \cdots, f_k\}$. Thus N_K denotes the number of feasible values of the relevant feature set $\{f_1, \cdots, f_K\}$ (size of the relevant state space). Also recall that $\mathbf{b_K}$ denotes a feasible instantiation of the relevant feature set $\{f_1, \cdots, f_K\}$.

Lemma 5.2.1. *For any $0 < \epsilon < 1$ and $0 < \delta < 1$ and $m_K = O(\frac{n^2 T^2}{\epsilon^2}$ $log(\frac{nN_K|A|}{\delta}))$, once all the $\mathbf{b_K}$'s have been visited m_K times, the $\hat{\pi}_{best}$ returned by FIND-MODEL(S) is based on at most the first K features from $\mathcal{F}$ and an $\frac{\epsilon}{T}$-approx of π_o with a high probability of at least $1 - 2\delta$.*

Thus it suffices to set m_k which defines when $\hat{\pi}_k$ stops predicting $\perp$ for a $\mathbf{b_k}$ as follows,

$$m_k = O(\frac{n^2 T^2}{\epsilon^2} log(\frac{nN_k|A|}{\delta})) \qquad (5.5)$$

Lemma 5.2.1 gives us the condition that needs to be satisfied to ensure that the $\hat{\pi}_{best}$ returned by FIND-MODEL(S) is an $\frac{\epsilon}{T}$-approx of π_o. However, it says nothing about how MLEM(S) should select its actions to ensure that this condition is satisfied. We conclude our analysis of MLEM(S) with a discussion of its action selection mechanism (Step M2) which ensures that the exploration condition from Lemma 5.2.1 holds.

5.2.4 Action Selection

In order to ensure that the condition of visits specified in Lemma 5.2.1 is met as quickly as possible, MLEM(S) uses the model-based RL algorithm RMAX [15]. We refer the reader to Section 4.1.4 for a brief account on how RMAX operates and what motivates us to use RMAX as our underlying RL algorithm.

Akin to MLES, MLEM(S) maintains a separate instance of RMAX for each of the possible $n + 1$ MDPs corresponding to the $n + 1$ possible models of π_o. At any iteration of MLEM(S), let the $\hat{\pi}_{best}$ returned by Step M1 be $\hat{\pi}_k$ and the MDP associated with it be $\mathcal{M}_k$. MLEM(S) then picks the stationary policy computed from the RMAX instance associated with $\mathcal{M}_k$ to decide on the next T step actions. The policy for the RMAX instance can be computed using any of the standard techniques, such as Value Iteration or Policy Iteration. MLEM(S) believes that k is the true value of K and hence attempts to explore all $\mathbf{b_k}$'s m_k times to satisfy the condition of visits from Lemma 5.2.1. The policy computed from the RMAX instance associated with $\mathcal{M}_k$ precisely helps it to achieve that. However, there is a possiblity that MLEM(S) might get stuck in a part of the state space where only some amongst the K relevant features are truly relevant (active). In that case, it might converge to exploiting based on a suboptimal model $\hat{\pi}_k$.

In order to avoid that, once in every $\lceil \frac{1-3\epsilon}{\epsilon} \rceil$ T-step iterations, MLEM(S) computes the policy slightly differently. First, it chooses a k randomly from 0 to n. The goal is to visit a new $\mathbf{b_k}$ which has not been visited m_k times. If such a visit is not possible, then exploit based on the current $\hat{\pi}_{best}$. The RMAX policy computation is then as follows. Assume that the state space of the underlying MDP comprise of all n features from $\mathcal{F}$. First, for all states of the MDP whose first k feature value (first k features from $\mathcal{F}$) have not been visited m_k times, provide them the exploratory bonus. For every other state use $\hat{\pi}_{best}$ to perform the Bellman back up. Note $\hat{\pi}_{best}$ only concerns itself with the features that it is based upon and not on all of the n features. Just like MLES, we call such a T-step planning iteration an *exploratory* iteration while the former a *greedy* iteration.

Now due to these exploratory iterations, $\hat{\pi}_K$ is chosen periodically as the random model in these exploratory iterations. Eventually by the implicit explore or exploit property of RMAX, it can be shown that at some exploratory iteration where MLEM(S) chooses $\hat{\pi}_K$ as the random model, it must achieve an expected return as high as $U^* - 2\epsilon$, with a high probability (since there are only finitely many entries to explore). Then from Assumption 1, we know that MLEM(S) must be following the optimal policy, otherwise such a high return would not have been possible. Thus MLEM(S) has learned a decent enough model of π_o that yields the optimal policy. Henceforth in every greedy iteration, it keeps exploiting based on this model and follows the optimal policy which eventually leads to a near optimal return. The details of how the above happens is exactly similar to the reasoning behind Lemma 4.1.2 of MLES, and is presented in Appendix A.3.

Lemma 5.2.2. *For any $0 < \epsilon < 1$ and $0 < \delta < 1$, with a high probability of at least $1 - 4\delta$, MLEM(S) achieves an actual return $\geq U^* - 5\epsilon$ against any Markovian o whose policy is based on the first K features from $\mathcal{F}$, in a number of time steps given by*

$$O(\frac{N_K n^3 T^3}{\epsilon^7} log(\frac{nN_K|A|}{\delta})log^2(\frac{1}{\delta})),$$

a quantity polynomial in $\frac{1}{\epsilon}$, $\frac{1}{\delta}$, n, N_K, $|A|$ and T.

Note our sample complexity argument handles the worst case scenario. Against most practical opponents, MLEM(S) will likely converge to near optimal behavior in far fewer samples.

The computational complexity of MLEM(S) for every planning iteration comprises two parts. The first part arises from FIND-MODEL(S), while the latter from the action selection step. FIND-MODEL(S) takes an order of $O(n^2)$ computations on each planning iteration. For the action selection step, we need to solve a MDP. Hence the computational complexity for this step is equivalent to that of any MDP solver, such as Value Iteration.

This concludes our discussion on MLEM(S). The goal of MLEM(S) was to solve the sequential structure learning problem, a simplification of the general problem. Next, we build on it to propose a more general subroutine which solves the same problem without the sequential structure assumption.

5.3 Model Learning and Exploitation of Markovian Agents without the Sequential Structure Assumption

In this section we present our subroutine for Step 2 on our road to specifying JOMA. As before we denote the agent under our control as i and the Markovian agents as a single agent o, with π_o being its unknown policy. Let $\mathcal{F}$ be the set of possible features that π_o might depend on and n be its cardinality. The only difference from the sequential structure learning problem is that now the ordering on the features in $\mathcal{F}$ may be arbitrary (based on the best possible guess) and not by relevance. In other words, there might be irrelevant features preceding the relevant features in $\mathcal{F}$.

Note, a very straightforward extension of MLEM(S) can solve the problem: arrange all features in some order and execute MLEM(S). The downside of this approach is that from Lemma 5.2.1, we now require sufficient visits to all of the K' feature values, where K' is the smallest value $\in [0, n]$ that spans over all relevant features. Observe that some of these K' features may be irrelevant. In many domains it is possible to guess a decent ordering of features and for such cases the above approach provides an efficient solution. However, in the worst case the last feature in the sequence may be a relevant feature and in that case MLEM(S) can only solve the problem by requiring sufficient visits to the entire state space determined from n features: the problem we intended to avoid in the very first place. Our goal in this section is to do better in terms of sample complexity. This goal motivates our new subroutine: *M*odel *L*earning and *E*xploitation against *M*arkovian agents, or the MLEM algorithm.

MLEM shares the same algorithmic structure as MLEM(S) (Algorithm 6) except for a crucial distinction pertaining to the algorithm used in Step M1 to compute $\hat{\pi}_{best}$.

5.3.1 FIND-MODEL-GENERAL Algorithm

Step M1 of MLEM is similar to that of MLEM(S) except now it makes a call to the FIND-MODEL-GENERAL algorithm. In a spirit similar to FIND-MODEL(S), the main objective of FIND-MODEL-GENERAL is to eventually return a $\hat{\pi}_{best}$ which is an $\frac{\epsilon}{T}$-approx of π_o with a high probability, by observing as few online samples of data as possible.

We begin with a brief intuitive explanation of how the FIND-MODEL-GENERAL algorithm operates. Unlike MLEM(S), MLEM has to maintain a model for every possible combination of features. From a high level perspective, FIND-MODEL-GENERAL chooses $\hat{\pi}_{best}$ by searching incrementally for the smallest model such that no larger model (ones based on greater number of features) is more predictive. In order to do so, it iterates over models ranging from the smallest to the largest. In the process of checking whether a model $\hat{\pi}$ is the best predictive model, FIND-MODEL-GENERAL compares it with all possible models that are based on the features from $\hat{\pi}$ and additional ones. If all the larger models fail to add to the predictiveness of π, then FIND-MODEL-GENERAL concludes that $\hat{\pi}$ is $\hat{\pi}_{best}$. If not, it moves on to the next model.

Algorithm 8 specifies the FIND-MODEL-GENERAL algorithm in detail. A brief summary of its key steps is as follows. Let $\mathcal{P}$ be the power set of the feature set $\mathcal{F}$ and $\mathcal{P}_k \subseteq \mathcal{P}$ be all the subsets of $\mathcal{P}$ with cardinality k. In other words, $\mathcal{P}_k$ is the set consisting of all combinations of k features from $\mathcal{F}$. The steps employed by FIND-MODEL-GENERAL are then as follows:

Lines 1-2. Iterate over $0 \leq k < n$ and for every k generate $\mathcal{P}_k$.

Line 3. Iterate over every element $\wp_k$ from this set, i.e., $\wp_k$ is a set of k features.

Lines 5-9: Let the model comprising the features in $\wp_k$ be $\hat{\pi}_{\wp_k}$. Now generate all possible sequences of size $k + 1$ to n, where the first k features in the sequence are the features from $\wp_k$.

Line 10: Execute FIND-MODEL(S) on each of these sequences.

Lines 11-18: If the model returned by FIND-MODEL(S) for all of these sequences is $\hat{\pi}_{\wp_k}$, then we are sure with a high probability that from the data samples seen so far, $\hat{\pi}_{\wp_k}$ is the best model (Lines 11-17). Return $\hat{\pi}_{\wp_k}$. If not, keep repeating the process until a particular $\hat{\pi}_{\wp_k}$ satisfies the condition. In the worst case, return the model comprising all features (Line 18).

Our main theoretical result concerning FIND-MODEL-GENERAL, analogous to the one for FIND-MODEL(S) (Lemma 5.2.1), is as follows. Let N'_{2K} be the maximum size of the state space from any $2K$-dimensional feature vector that can be selected from the n features.

Lemma 5.3.1. *Once all the feasible values of any $2K$-dimensional feature vector originating from possible combinations of $2K$ features from $\mathcal{F}$ are visited $m'_{2K} = O(\frac{n^3 T^2}{\epsilon^2} log(\frac{n N'_{2K} |A|}{\delta}))$ times, then the $\hat{\pi}_{best}$ returned by FIND-MODEL-GENERAL is an $\frac{\epsilon}{T}$-approx of π_o, with a high probability of at least $1 - 2\delta$.*

The proof is presented in Appendix B.1. Lemma 5.3.1 states the sufficient condition that needs to be satisfied to ensure that the $\hat{\pi}_{best}$ returned by

---

Algorithm 8. FIND-MODEL-GENERAL

```
 1 for 0 ≤ k < n do
 2 │   Generate 𝒫ₖ ;
 3 │   for every ℘ₖ ∈ 𝒫ₖ do
 4 │   │   flag ← true;
 5 │   │   for 1 ≤ k' <= n − k do
 6 │   │   │   Generate 𝒫ₖ';
 7 │   │   │   for every ℘ₖ' ∈ 𝒫ₖ' do
 8 │   │   │   │   if ℘ₖ' ∩ ℘ₖ ≠ φ then
 9 │   │   │   │   └   continue;
 9 │   │   │   │   seq ← a sequence starting with ℘ₖ followed by ℘ₖ';
10 │   │   │   │   π̂ ← execute FIND-MODEL(S) on seq;
11 │   │   │   │   if π̂ ≠ π̂℘ₖ then
12 │   │   │   │   │   flag ← false;
13 │   │   │   │   └   break;
14 │   │   │   if !flag then
15 │   │   │   └   break;
16 │   │   if flag then
17 │   │   └   return π̂℘ₖ
18 return model comprised of all features
```

FIND-MODEL-GENERAL suffices to be an $\frac{\epsilon}{T}$-approx of π_o with a high likelihood. Analogous to Equation 5.5, for any arbitrary $k \in [0, n]$, the m'_k term is computed as follows:

$$m'_k = O(\frac{n^3 T^2}{\epsilon^2} log(\frac{n N'_k |A|}{\delta})) \qquad (5.6)$$

To understand the condition specified in Lemma 5.3.1, consider the following example. Let the feature set be $\{X_1, X_2, X_3, X_4, X_5\}$ and let each feature assume binary values, i.e. $X_i \in \{0, 1\}$. Assume that the relevant features are X_1 and X_3. So $n = 5$ and $K = 2$. Then all 4-dimensional feature vectors consist of the following vectors: $\{X_1, X_2, X_3, X_4\}$, $\{X_1, X_2, X_3, X_5\}$, $\{X_2, X_3, X_4, X_5\}$, $\{X_1, X_2, X_4, X_5\}$ and $\{X_1, X_3, X_4, X_5\}$: in total $\binom{5}{4} = 5$ combinations. N'_4 is the maximum size of the state space from any of these 4-dimensional feature vectors. In this case $N'_4 = 2^4 = 16$. Then the condition in Lemma 5.3.1 requires all possible values of these 4-dimensional feature vectors ($2^4 \times 5 = 80$ of them) to be visited m'_4 times, where m'_4 is computed using Equation 5.6.

Note that the condition of visits specified in Lemma 5.3.1 is looser than that from Lemma 5.2.1 (for MLEM(S)). Lemma 5.2.1 required sufficient visits to only the relevant state space, whereas Lemma 5.3.1 requires sufficient visits to certain parts of state space that are irrelevant. Namely it requires

visits to all values of any $2K$-dimensional feature vector. This extra exploration is necessary to figure out the correct features for π_o and can be seen as
the price paid for not having any prior knowledge of π_o. Our detailed proof
of Lemma 5.3.1 in Appendix B.1 provides a formal justification of how we
arrived at this condition.

Lemma 5.3.1 gives us the condition that needs to be satisfied to ensure
that the $\hat{\pi}_{best}$ returned by FIND-MODEL-GENERAL is an $\frac{\epsilon}{T}$-approx of π_o.
It says nothing about how the condition is satisfied. This is assured by the
action selection mechanism of MLEM.

5.3.2 Action Selection

The onus lies on the action selection mechanism to ensure that the condition
of visits specified in Lemma 5.3.1 is met as quickly as possible which will allow
FIND-MODEL-GENERAL to keep returning a near accurate $\hat{\pi}_{best}$ consistently.

In spirit similar to MLEM(S), MLEM now maintains an RMAX instance
for each of the 2^n possible MDPs, i.e., MDPs originating from all possible
combinations of features as the state space. Also akin to MLEM(S), MLEM
interleaves between greedy and exploratory iterations. At the beginning of
each T-step greedy iteration, MLEM picks the RMAX instance associated
with the MDP comprised of features from $\hat{\pi}_{best}$ and uses it to compute the
policy to follow for the next T steps.

Whereas at the beginning of each T-step exploratory iteration, MLEM
chooses a k randomly from 0 to n. The policy computation is similar to the
one for a greedy iteration with a couple of key differences. Assume that the
state space of the underlying MDP comprise all n features from $\mathcal{F}$. First, for
all states of the MDP which have some k feature vector value that has not
been visited m'_k times, it provides them the exploratory bonus. For every
other state it uses $\hat{\pi}_{best}$ to perform the Bellman back up. Once again note
that the $\hat{\pi}_{best}$ only concerns itself with the features on which it is based and
not on all of the n features. The objective of the exploratory iteration is to
facilitate the exploration needed to satisfy the condition in Lemma 5.3.1.

Then a sample complexity analysis analogous to that of Lemma 5.2.2,
brings us to our main theoretical result concerning MLEM. The term $\mathbf{N_{2K}}$
in the statement of the Lemma is the total number of feasible values of all $2K$-
dimensional feature vectors (originating from all combinations of $2K$ features
from $\mathcal{F}$). All other terms have their usual meaning from Lemma 5.2.2.

Lemma 5.3.2. *For any $0 < \epsilon < 1$ and $0 < \delta < 1$, with a high probability
of at least $1 - 4\delta$, MLEM achieves an actual return $\geq U^* - 5\epsilon$ against any
Markovian o which uses features from $\mathcal{F}$, in a number of time steps given by*

$$O(\frac{\mathbf{N_{2K}} n^4 T^3}{\epsilon^7} log(\frac{n N'_{2K} |A|}{\delta}) log^2(\frac{1}{\delta})),$$

a quantity polynomial in $\frac{1}{\epsilon}$, $\frac{1}{\delta}$, n, $\mathbf{N_{2K}}$, N'_{2K}, $|A|$ and T.

The bound from Lemma 5.3.2 is similar to that of Lemma 5.2.2, except for the obvious exceptions that the number of entries we need to explore now is $\mathbf{N_{2K}}$ and we need to visit each such entry m'_{2K} times (as opposed to N_K and m_K for Lemma 5.2.2). The proof is also similar to that for Lemma 5.2.2 and hence we skip it. We acknowledge that our sample complexity bound in this case is worse than the sample complexity bound proven for the sequential structure learning problem (namely Lemma 5.2.2) as it is polynomially dependent on $\mathbf{N_{2K}}$, instead of the size of the relevant state space. As suggested earlier, the weakening of the bound can be seen as the price we pay for not having any prior information about π_o. However in cases where $\mathbf{N_{2K}}$ is significantly smaller than the state space from the entire $\mathcal{F}$, the savings in terms of sample complexity is still very significant.

Finally, some comments on the computational complexity of MLEM. FIND-MODEL-GENERAL takes an order of $O(n2^n)$ computations on each planning iteration. However most of these computations can be done in parallel if the underlying hardware supports parallelism. For the action selection step, once again the computational complexity is that of any MDP solver.

That completes our analysis of how MLEM solves the general structure learning problems. In the following section, we present the full blown JOMA algorithm.

5.4   Joint Optimization against Markovian Agents

Until now, we have focused on the scenario in which there is just one MLEM (or, MLEM(S)) agent in the population. Now, we account for the case where there is more than one JOMA agent in the population in addition to the Markovian agents. The JOMA agents know each other's identities and are only concerned with modeling the Markovian agents in the population.

Since the JOMA agents individually prefer different joint-returns based on their own utility functions, we require them to agree on one that maximizes social welfare (SW). It is important that the JOMA agents agree on a common ordering between themselves and their actions. For example consider the special case of self-play between two JOMA agents playing Battle of Sexes (Table 5.1). This is a very special case pertaining to a situation when the number of Markovian agents in the population is 0. There are two outcomes in the game that maximize SW. If the agents are uncoordinated, they may target different outcomes and end up playing different actions leading to a SW value of 0. However if they agree on a unique ordering of themselves and their actions, they can use a deterministic rule to coordinate and attain one of the SW maximizing outcomes (like iterating over all agents and their actions in the manner prescribed by their ordering and choosing the first joint-action that maximizes SW).

We assume that the game is defined in a canonical form so that all agents see it defined in exactly the same way, in particular including the same

Table 5.1 Payoff matrix for Battle of Sexes (BoS)

	a1	a2
a1	(1/3,2/3)	(0,0)
a2	(0,0)	(2/3,1/3)

ordering over actions. So an ordering over the actions is implicit. However the game definition does not include a specification of which agents are to play the game, so does not (and can not) specify an ordering over the agents. Such an ordering must be determined either by convention (e.g. ascending order of MAC id) or by the agents themselves. If a convention is unavailable, we require the JOMA agents to settle on a common shared ordering between themselves in a distributed fashion.

Henceforth we denote the number of JOMA agents in the population by J. Also keeping in line with our past analysis we assume that the maximum SW value is bounded by $R_{max} = 1$. JOMA's functionality can be broken down into two steps.

Step 1. First JOMA strives to settle on a common ordering amongst all the self agents in a distributed fashion. We call this phase of JOMA the *ordering computation* phase. Assume there are 5 JOMA agents in the population. Without loss of generality, we refer to them as A, B, C, D and E from the perspective of the JOMA agent whom we are representing in this computation (that is the naming is arbitrary and may differ for all the JOMA agents). Recall that we have a predefined ordering among the actions for every agent. Thus for every agent we can label the actions as action $1, 2, \ldots$.

As part of the ordering computation phase, each JOMA agent plays randomly for $T1 = O(Jlog(\frac{J}{\delta}))$ time steps. We later specify the rationale behind the value of $T1$. Initially JOMA groups all the 5 agents in a single set (refer to Figure 5.1). Now whenever on a particular time step an agent plays a different action from its peers, it splits this set. For example in Figure 5.1, at some time step agents A and B play action 1, agent C plays action 3, while agents D and E play action 4. Then JOMA splits this set into three sets and defines an ordering among them pertaining to the different actions played. In our example this leads to three sets: $S_1 = \{A, B\}$, $S_2 = \{C\}$ and $S_3 = \{D, E\}$, with S_1 preceding S_2 (denoted by $S_1 < S_2$) and S_2 preceding S_3 (denoted by $S_2 < S_3$). S_1 precedes S_2 because agents from S_1 played action 1 while agents from S_2 played action 3, and action 1 precedes action 3 in the ordering of actions. This also means that in the final ordering elements from S_1 would always precede that of S_2 and elements from S_2 would always precede that of S_3.

JOMA continues this process of splitting sets until either of the following two events happen: (1) it reaches the desired split of J singleton sets with each set containing an agent; (2) $T1$ time steps elapse. We later show as part of proof of Theorem 5.4.1 that through our choice of $T1$, we ensure

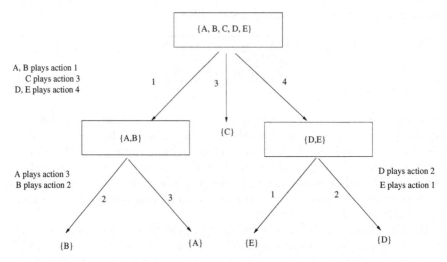

Fig. 5.1 Example of how the JOMA agents agree on an ordering. The final ordering can be obtained by reading all the sets from the leaf nodes from left to right, i.e., $B < A < C < E < D$.

that through $T1$ steps of random play by the JOMA agents, we always reach the desired split of J singleton sets with an error probability of at most δ. This in turn means that the JOMA agents settle on a common shared ordering among themselves with an error probability of at most δ.

Step 2. Once settled on a common ordering, JOMA switches to following MLEM (or MLEM(S) depending on the nature of the problem). Each JOMA agent runs MLEM independently. Since they are solving the same problem and have agreed on a common ordering between themselves (the ordering over actions being implicit), they arrive at the same solution, i.e., same stationary joint-policy to follow for the ensuing T steps. Then they follow their share of the joint-policy for the next T steps.

Next, we state and prove the main theoretical property of JOMA.

Theorem 5.4.1. *Consider a population of JOMA and Markovian agents (at least one) which use features from $\mathcal{F}$. Then for any $0 < \epsilon < 1$ and $0 < \delta < 1$, with a high probability of at least $1 - 5\delta$, the JOMA agents achieve an actual joint-return at least within 6ϵ of the SW maximizing expected joint-return, in a number of time steps given by:*

$$O(\frac{J}{\epsilon}log(\frac{J}{\delta}) + \frac{N_K n^3 T^3}{\epsilon^7}log(\frac{nN_K|A|}{\delta})log^2(\frac{1}{\delta}))$$

if the problem can be formulated as a sequential structure learning problem, otherwise

$$O(\frac{J}{\epsilon}log(\frac{J}{\delta}) + \frac{\mathbf{N_{2K}}n^4T^3}{\epsilon^7}log(\frac{nN'_{2K}|A|}{\delta})log^2(\frac{1}{\delta}))$$

The first quantity is polynomial in $\frac{1}{\epsilon}$, $\frac{1}{\delta}$, J, n, N_K, $|A|$ and T, while the latter in $\frac{1}{\epsilon}$, $\frac{1}{\delta}$, J, n, $\mathbf{N_{2K}}$, N'_{2K}, $|A|$ and T.

Proof. The proof follows in two parts.

First, lets prove that all the JOMA agents settle on the same ordering with a high likelihood. Note that the number of splits in the tree from our ordering algorithm (Figure 5.1) can be at most $J-1$. The probability of any such split depends on the probability of any pair of JOMA agents from that node of the tree choosing different actions.

For a pair of JOMA agents, with at least two actions, the probability that they would select a different action on a time step by playing randomly is at least $\frac{1}{2}$. So the expected number of time steps taken by them to choose a different action is at most 2. Then by Hoeffding bound [40], a node which contains more than one JOMA agent splits in $O(2log(\frac{J}{\delta})) \equiv O(log(\frac{J}{\delta}))$ time steps with error probability at most $\frac{\delta}{J}$. Since there can be at most $J-1$ splits to consider, then in $T1 = O(Jlog(\frac{J}{\delta}))$ time steps the JOMA agents settle on a unique ordering among them with an error probability at most δ (from union bound).

Second, from Lemma 5.3.2 we know that once the JOMA agents have settled on a common shared unique ordering, JOMA achieves an actual joint-return within 5ϵ of the SW maximizing expected joint-return, with a probability of at least $1 - 4\delta$ in $T2$ time steps, where

$$T2 = O(\frac{N_Kn^3T^3}{\epsilon^7}log(\frac{nN_K|A|}{\delta})log^2(\frac{1}{\delta})), \text{ for the seq structure learning prob;}$$

$$= O(\frac{\mathbf{N_{2K}}n^4T^3}{\epsilon^7}log(\frac{nN'_{2K}|A|}{\delta})log^2(\frac{1}{\delta})) \text{ otherwise.} \qquad (5.7)$$

Also since we have to compensate for the poor return over the initial $T1$ time steps, we need to run JOMA for an additional $\frac{T1}{\epsilon}$ time steps to compensate and achieve an actual joint-return within 6ϵ of the SW maximizing expected joint-return. Then combining the claims from all parts of the proof using union bound and observing that the total sample complexity is $\frac{T1}{\epsilon}+T2$, gives us the proof. $\qquad\square$

This concludes our theoretical analysis of JOMA catering to the case when the JOMA agents know each others' identities to begin with. In the following section we address the case when the JOMA agents are unaware of each others' identities and has to detect one another as part of the learning process.

5.5   JOMA with Identities Unknown

We retain the same algorithmic outline of JOMA. Only our analysis differs to address this general case.

We first show that at the end of the ordering computation phase, if two or more agents end up being clubbed in the same node, then the probability of one of them being a JOMA agent is as low as δ. Each JOMA agent plays randomly in the ordering computation phase. Assume at any step of this phase, a non-JOMA agent plays action a. The probability that a JOMA agent plays the same action is at most $1/2$ (since the action space includes at least 2 actions for the JOMA agents). So the expected number of time steps taken by a JOMA agent to play a different action from a non-JOMA agent is at most 2. Hence through our choice of the length of the ordering computation phase and by a similar line of reasoning as used in the proof of Theorem 5.4.1, it can be shown that the probability of a JOMA agent sharing a node with another agent at the end of the ordering computation phase is as low as δ. So if at the end of the ordering computation phase, two agents share a same node, we can be sure with a high probability of $1 - \delta$, that these agents are non-JOMA agents. Henceforth, JOMA treats these agents as non-JOMA agents and switches to following MLEM (or MLEM(S) depending on the nature of the problem).

However, what if there is a non-JOMA agent that does not share a node with any other agent at the end of the ordering computation phase? Clearly this agent has evaded detection. In that case the JOMA agents continue treating is as being another JOMA agent unless it plays a different action other than the one prescribed by MLEM (or MLEM(S)). JOMA successfully identifies it as a non-JOMA agent and restarts the MLEM subroutine. However in the very unlikely scenario if these other agents keep playing the action prescribed for them by JOMA forever, then JOMA ends up treating them as a self agent forever. It then successfully converges to the SW maximizing joint-return, where the SW value is computed as sum of returns of not only the JOMA agents, but also the ones who pretend to be JOMA. If not, then eventually once all non-JOMA agents are detected, then in an additional number of time steps (from Theorem 5.4.1), JOMA converges to achieving its desired SW maximizing joint-return.

This concludes our complete theoretical analysis of JOMA.

5.6   Empirical Validation

Whereas the main contribution of this paper is the introduction of JOMA as a theoretically grounded MAL algorithm, we would also like for it to be useful in practice. In this section we present some results of relatively simple experiments testing how JOMA performs in the presence of Markovian and non-Markovian agents. We begin by specifying our implementation of JOMA.

Theoretically, the specification of JOMA depends on the following input parameters: δ, ϵ, T and $\mathcal{F}$. These inputs determine the m_k for each model. Recall that m_k is the number of visits we require to each $\mathbf{b_k}$ to consider the estimate $M_k(\mathbf{b_k})$ (empirical distribution of o's play for $\mathbf{b_k}$) reliable. Furthermore, we require the (ϵ, T) pair to satisfy Assumption 1.

An implementation of JOMA straight from its theoretical specification is challenging for the following two reasons.

1. First, there exists no principled way of guessing an (ϵ, T) pair a priori that satisfies Assumption 1.
2. Second, even if we know such an (ϵ, T) pair, the value of each m_k computed based on it is prohibitively high for practical purposes. Note, the definition of m_k is a very conservative one and in practice much smaller values of m_k should suffice.

Hence we introduce a few approximations when implementing JOMA. First, instead of seeding JOMA with a δ, ϵ, $\mathcal{F}$ and T, we seed it with an m, δ and $\mathcal{F}$. m plays the role of m_k and is the same for models of all sizes. δ is required to compute the value of σ_k (Equation A.3). All our results are reported for $m = 20$ and $\delta = 0.2$. $\mathcal{F}$ in our case consists of the past 8 joint-actions. Each joint-action is treated as a feature. In other words, we let JOMA figure out which amongst these past 8 joint-actions can be best used to model each of the opponent algorithms. We run JOMA using MLEM(S) as the default fall back algorithm for Step 2.

Also, note that MLEM(S) needs an exploratory iteration once every $\lceil \frac{1-3\epsilon}{\epsilon} \rceil$ planning iterations. Since we do not specify a value for ϵ, it is not clear when to opt for an exploratory iteration. Hence we opt against an exploratory iteration. In all of our experiments, the explorations that happen in the greedy iterations are sufficient to generate good results.

Finally, MLEM(S) functions by planning for T time steps at a time. Such a T-step action selection policy is just the stationary RMAX policy computed by running Value Iteration in the underlying induced MDP and executed for T steps. In our implementation, we keep executing the computed stationary RMAX policy forever, unless a new state of the underlying induced MDP gets visited for the m'th time. In that case, we recompute it. This approach is structurally similar to the one described in Algorithm 6, except that it is more computationally efficient.

Our experiments involve empirical testing in selected multi-player games from gamut.[5] The MAL algorithms chosen as benchmarks for comparison are Cautious Fictitious Play [17], AWESOME [27] and GIGA-WOLF [12]. Each of these algorithms has its own specific objectives and is a popular representative from its own family of MAL algorithms. Cautious Fictitious Play represents the class of Fictitious Play algorithms that achieve universal consistency. AWESOME represents the family of MAL algorithms that achieve convergence and rationality in arbitrary repeated games. Meanwhile,

---

[5] http://gamut.stanford.edu

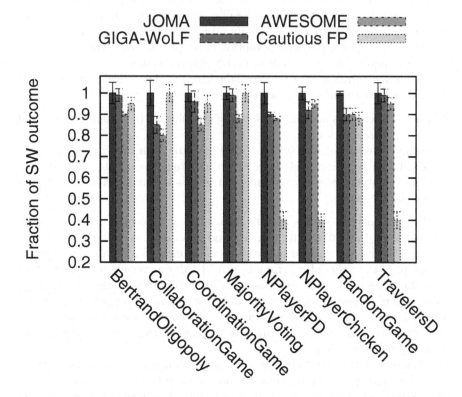

Fig. 5.2 Comparative results in different games from gamut. The experiments were conducted over 3 player versions of each game, with two self agents and the third agent drawn randomly from the set of benchmark agents, along with minimax, self agent and arbitrary memory-bounded strategies of size 3. The X axis of the plot shows the different games while the Y axis shows the converged SW value for each algorithm achieved in a game as a fraction of the best converged SW value achieved by any algorithm for that game.

GIGA-WOLF represents the family that achieves convergence and no-regret in repeated games.

Figure 5.6 presents our results. The experiments were conducted over 3 player versions of each game, with two self agents and the third agent drawn randomly from the set of benchmark agents, along with minimax, self agent and arbitrary memory-bounded strategies of size 3.[6] The X axis of the plot shows the different games while the Y axis shows the converged SW value for each algorithm achieved in a game as a fraction of the best converged SW value achieved by any algorithm for that game. Each point in the plot

---

[6] Arbitrary memory bounded agents of memory size 3 generated randomly.

is an average over 100 such plays. JOMA consistently achieves the highest SW in each of the games beating the next best algorithm by a significant margin in many games. The difference with the next best is statistically significant except in Bertrand Oligopoly, Collaboration Game, Coordination Game, Majority Voting and Travelers Dilemma (p-values < 0.05 from T-test in all other games).

Note, the only agents from the above set for which JOMA provides any theoretical guarantee are self-play, minimax and the adaptive strategy (based on features from $\mathcal{F}$). JOMA successfully models minimax as a memoryless opponent, while the adaptive strategy as a memory size 3 opponent. Though the MAL algorithms are non Markovian from JOMA's perspective, it still exploits them well. In all of our experiments involving the MAL algorithms, JOMA successfully models them based on at most the past 4 joint-actions from history.

5.7 Summary

This chapter introduces a novel MAL algorithm, called JOMA, that in an arbitrary repeated matrix game, in the presence of Markovian agents in the population, is the first to provably achieve a joint-return very close to the SW maximizing joint-return. Along with a thorough theoretical analysis of JOMA's sample complexity properties, we also present some initial empirical results from the gamut testbed demonstrating its relative effectiveness compared to some of its predecessors. In the next chapter we present more simpler variants of the MLEM and MLEM(S) subroutines, which are easier to implement. Though these new algorithms fail to provide similar theoretical guarantees of optimality as MLEM and MLEM(S), they seem to work well in practice.

Chapter 6
Targeted Modeling of Markovian Agents

In the previous chapter, we introduced an algorithm JOMA that in the presence of Markovian agents in the population, provably achieves a joint return very close to the SW maximizing joint return by exploiting the Markovian agents maximally, in efficient sample complexity. We assume that JOMA has some prior knowledge of the possible set of features $\mathcal{F}$ upon which the Markovian agents may base their policies, but not the exact set. Being unaware of the exact feature set that determines the Markovian agents' policies, JOMA maintains different models of the latter based on different combinations of features from $\mathcal{F}$. Then in online repeated play, the main objective of JOMA is to figure out which amongst these models successfully determine the policies of the Markovian agents and consequently plan its actions based on that. In order to do so, JOMA relies on two subroutines, namely MLeM(S) and MLeM, that implement its two main components, namely the model selection and the action selection component.

However as observed in Section 5.6, an implementation of these subroutines straight from their theoretical specification is challenging. We refer the reader to Section 5.6 for a complete account of why that is so. Our main goal in this chapter is to propose simpler variants of these subroutines that are easier to implement. In the process we forego the neat theoretical properties of MLeM(S) and MLeM, however based on our systematic empirical analysis presented in this chapter, the proposed approaches appear to work well in practice.

Just like MLeM (and hence MLeM(S)), our setting of interest in this chapter is a repeated game between an agent under our control and other Markovian agents. Once again observing that we can treat all the Markovian agents in the population as a single Markovian agent whose policy and action space is just the joint policy and joint action space of all the Markovian agents respectively, we treat the problem as a two player repeated game against a Markovian agent o with an unknown policy π_o and action space A. We assume prior knowledge of a set of features $\mathcal{F}$ which is of size n, some of which are assumed to characterize the unknown π_o. The purpose of this chapter is then

D. Chakraborty, *Sample Efficient Multiagent Learning in the Presence*   73
of Markovian Agents, Studies in Computational Intelligence 523,
DOI: 10.1007/978-3-319-02606-0_6, © Springer International Publishing Switzerland 2014

to propose a couple of algorithms, called *Targeted Opponent Modeler for Markovian Agents* (or TOMMA) and its sequential counterpart TOMMA(S), that suffice as simpler variants of MLEM and MLEM(S) respectively.

The remainder of this chapter is organized as follows. Section 6.1 introduces TOMMA(S) and TOMMA, Sections 6.2 and 6.3 introduce a couple of surveillance based domains used as test beds for our empirical analysis along with some empirical results, and Section 6.4 summarizes the chapter.

6.1 Algorithms and Analysis

Just like we did for MLEM, we begin by proposing a solution to a simplified version of the modeling problem that relies on the sequential structure assumption: the *sequential structure learning problem* (see Definition 21 from Chapter 5). Here the features in $\mathcal{F}$ are arranged in a sequence such that all of the K relevant features that determine π_o precede the irrelevant ones in the sequence, with K being unknown. We call our variation of TOMMA for this problem TOMMA(S), with the S standing for the sequential structure assumption. Later we build on it to solve the general problem where $\mathcal{F}$ does not satisfy the sequential structure assumption, and present TOMMA.

6.1.1 TOMMA(S)

The concept of a model for π_o for TOMMA(S) is very similar to that for MLEM(S) except for one crucial difference. For the sake of completeness, we reintroduce the concept of a model and in the process explain this difference.

Since TOMMA(S) is unaware of the exact K that characterizes π_o, it maintains a model of π_o for each set of features that can be incrementally generated by choosing the first k features from $\mathcal{F}$, $k \in [0 : n]$. Let the model that is based on the first k features be $\hat{\pi}_k$. Formally, $\hat{\pi}_k : \{f_1, \ldots, f_k\} \mapsto \Delta A$. Note $\hat{\pi}_0$ is a model that is completely stationary, and independent of any features. Internally each $\hat{\pi}_k$ maintains a value $M_k(\mathbf{b_k})$ which is the maximum likelihood distribution of o's play, for every possible value $\mathbf{b_k}$ of the first k features from $\mathcal{F}$. Whenever the first k features assume a value $\mathbf{b_k}$ in online play, we say a *visit* to $\mathbf{b_k}$ has occurred. $\hat{\pi}_k(\mathbf{b_k})$ is then defined as,

$$\hat{\pi}_k(\mathbf{b_k}) = \begin{cases} M_k(\mathbf{b_k}) & \text{once } visit(\mathbf{b_k}) = m, \\ \perp & \text{when } visit(\mathbf{b_k}) < m; \end{cases} \quad (6.1)$$

where $visit(\mathbf{b_k})$ is the number of times $\mathbf{b_k}$ has been visited, and m is a system level parameter. In other words, once a $\mathbf{b_k}$ is visited m times, we consider the estimate $M_k(\mathbf{b_k})$ reliable, and assign $\hat{\pi}_k(\mathbf{b_k})$ to it. Henceforth we make no updates to $\hat{\pi}_k(\mathbf{b_k})$ (for $visit(\mathbf{b_k}) > m$). If a reliable estimate of $M_k(\mathbf{b_k})$ is unavailable (when $visit(\mathbf{b_k}) < m$), then $\hat{\pi}_k(\mathbf{b_k})$ is set to $\perp$ (meaning "I do not know").

Algorithm 9. TOMMA(S) AND TOMMA

---

input: $\mathcal{F}$, m, δ, T

;

repeat

    Determine $\hat{\pi}_{best}$ (best predictive model of π_o);;

    Compute T-step action selection policy using $\hat{\pi}_{best}$;;

    $\tau \leftarrow 1$;;

    repeat

        Execute the action selection policy;;

        $\tau \leftarrow \tau + 1$;;

    until $\tau > T$;

    Update all models based on past T joint actions;;

until *forever*;

Note the main difference with the definition of a model pertaining to MLEM(S) (Equation 5.1) is that now we require the same number of visits to all $\mathbf{b_k}$'s to consider the estimate $M_k(\mathbf{b_k})$ reliable, i.e., $\forall k, m_k = m$.

Having introduced the concept of a model, we next present the algorithmic outline for TOMMA(S) (Algorithm 9). The inputs to Algorithm 9 are the feature set $\mathcal{F}$, m, a small probability value δ and the planning horizon T. Algorithm 9 operates by planning for T time steps at a time. At the beginning of every such *planning iteration*, it computes a best estimate model for π_o, denoted by $\hat{\pi}_{best}$, based on its past interactions with o. It then uses $\hat{\pi}_{best}$ to compute a T-step action selection policy which it follows for the next T steps. Both TOMMA(S) and TOMMA share the same algorithmic outline of Algorithm 9. The two places in Algorithm 9 where they differ are (i) how they compute their respective $\hat{\pi}_{best}$, and (ii) how they compute the T-step action selection policy. We first show how TOMMA(S) addresses these two sub-problems.

Model Selection: The objective of the model selection component of TOMMA(S) is to output the best predictive model for π_o from all of the $n + 1$ models maintained. We call a model to be of *size k* if it uses the first k features from $\mathcal{F}$. Initially $\hat{\pi}_0$ is assigned to $\hat{\pi}_{best}$. Recall that the first K features from $\mathcal{F}$ completely determine π_o, K being unknown. Then, all models of size $\geq K$ can learn π_o accurately (as they include all of the relevant features), with the bigger models requiring more samples to do so. On the other hand, models of size $< K$ cannot fully represent π_o. Then TOMMA(S) chooses $\hat{\pi}_{best}$ by comparing models of increasing size, to determine the shortest most descriptive model such that the next larger model ceases to be more predictive of π_o.

A careful reader will notice that there is a significant difference between the model selection mechanism of TOMMA(S) and MLEM(S), though they both appear to be very similar. In case of MLEM(S), we search for the shortest model such that all larger models cease to be more predictive, not just the next larger one.

Assume at some planning iteration, $\hat{\pi}_{best} = \hat{\pi}_k$. In the next planning iteration, TOMMA(S) compares $\hat{\pi}_k$ with $\hat{\pi}_{k+1}$. In order to do so, it first computes a Δ_k that is the difference in prediction of $\hat{\pi}_k$ and $\hat{\pi}_{k+1}$ (computed using Equation 5.3). Then it computes a σ_k. If $\Delta_k = -1$, then we assign $\sigma_k = 1$. If $\Delta_k \neq -1$, then the computed σ_k is the tightest estimate satisfying the following condition (akin to the computation of σ_k for MLEM(S), see Equation 5.4):

$$\Pr(\Delta_k < \sigma_k) > 1 - \delta \text{ if } \mathbb{E}(\Delta_k) = 0 \tag{6.2}$$

where δ is the small probability value taken as input. By tightest we mean an estimate as close to Δ_k as possible. What Equation 6.2 suggests is that given the data seen so far, if there is no statistical evidence to the claim that a model spanning over the first $k + 1$ features from $\mathcal{F}$ is more predictive than one spanning over the first k features from $\mathcal{F}$ (i.e. $\mathbb{E}(\Delta_k) = 0$), then Δ_k should always be less than the computed σ_k with a high probability of at least $1 - \delta$. A careful reader would observe that for $K \leq k < n$, $\mathbb{E}(\Delta_k)$ is always 0. However in the initial stages of learning, $\mathbb{E}(\Delta_k) = 0$ may also hold for some $k < K$. This is generally true when we have not explored enough to deduce the relevance of all K features.

In such cases σ_k is assigned as follows:

$$\sigma_k = \sqrt{\frac{1}{m} log(\frac{2|A|N_{k+1}}{\delta})} \tag{6.3}$$

Recall from Chapter 5 that N_{k+1} is the size of the feature space $\{f_1, \ldots, f_{k+1}\}$. It can be shown through a couple of applications of Hoeffding bound and union bound, that if we assign σ_k using Equation 6.3, then the condition from Equation 6.2 remains satisfied. Complete details of how we arrived at this are presented in Appendix C.1.

Now if Δ_k exceeds σ_k, TOMMA(S) has reason to believe that $\hat{\pi}_k$ is not the correct model for π_o and it updates $\hat{\pi}_{best}$ to $\hat{\pi}_{k+1}$; otherwise it retains $\hat{\pi}_k$ as $\hat{\pi}_{best}$. Later at the end of this Section, we explain that once converged to the correct model $\hat{\pi}_K$, the error probability with which TOMMA(S) switches mistakenly to a bigger model is upper-bounded by δ.

Note that our model selection in this case is much greedier than that of MLEM(S). We could have used the same model selection component of MLEM(S) (i.e., FIND-MODEL(S)) over here as well. However, we want to deliberately use a greedier model selection to pave way for the model selection of TOMMA, which is greedy in similar fashion and computationally more efficient than that of MLEM.

Action Selection: Akin to MLEM(S), the T-step action selection policy for TOMMA(S) is based on RMAX [15]. Now, assume at some planning iteration,

$\hat{\pi}_{best} = \hat{\pi}_k$, which means that data from past plays suggest that $\hat{\pi}_k$ is as predictive as $\hat{\pi}_{k+1}$, with a high likelihood. To be more certain about this hypothesis, TOMMA(S) strives to explore the entire feature space pertaining to the feature set $\{f_1, \ldots, f_{k+1}\}$, m times. In order to do so, it follows RMAX assuming the state space is determined by the feature space $\{f_1, \ldots, f_{k+1}\}$, while the transition and the reward functions are determined by $\hat{\pi}_{k+1}$. Two things can happen from there onwards. Either, (1) because of this exhaustive exploration of the feature space $\{f_1, \ldots, f_{k+1}\}$, it infers that $\hat{\pi}_{k+1}$ is indeed more predictive than $\hat{\pi}_k$, and switches to $\hat{\pi}_{k+1}$ as $\hat{\pi}_{best}$. Or, (2) the above does not happen, and RMAX converges to exploiting based on $\hat{\pi}_{k+1}$. The hope is that by following this incremental style of exploration, it will incrementally switch through different models, until it converges to $\hat{\pi}_K$. From that point onwards it never switches to a bigger model, with a high likelihood of $1 - \delta$.

However unlike MLEM(S), there remains a chance that TOMMA(S) may get stuck at a local optimum by converging to a smaller sized model because of insufficient exploration. This generally happens when the exploration is restricted to only a part of the state space, where only some amongst the relevant features are truly active. We consider this to be an acceptable tradeoff, especially in time critical missions, where the goal is to quickly compute a reasonable model of the other agent, and act based on it.

This concludes our presentation of TOMMA(S). We next build on it to propose the full blown TOMMA algorithm that does not make the sequential structure assumption for $\mathcal{F}$.

6.1.2 TOMMA

The most important difference between the general modeling problem, and its sequential counterpart is that the former does not have access to a feature set $\mathcal{F}$ with the relevant features preceding the irrelevant ones. So unlike TOMMA(S), TOMMA does not have access to an ordered model space which it can incrementally search for the correct model.

TOMMA maintains a model for every combination of features from $\mathcal{F}$, and it sorts them in a sequence such that the ones which are *quicker to learn* precede the others. For example if the features are based on different memory sizes (as in our empirical analyses in Sections 6.2 and 6.3), models which are based on features computed from smaller memory sizes precede the ones based on features computed from bigger memory sizes since they are quicker to learn. If such a predefined ordering is not feasible, then we proceed with an arbitrary ordering of models sorted based on their sizes. TOMMA then incrementally searches this model space to find the first model from this sequence that determines π_o.

Algorithm 10 gives an example of a comparator function that can be used to compare two models $\hat{\pi}$ and $\bar{\pi}$ which are based on features computed from different memory sizes. The function $MemSize(\hat{\pi}, x)$ first arranges the

---

Algorithm 10. A SAMPLE MODEL COMPARATOR FOR TOMMA

---

input: $\hat{\pi}, \bar{\pi}$
;
$s_1 \leftarrow$ size of $\hat{\pi}$; $s_2 \leftarrow$ size of $\bar{\pi}$;;
while $s_1 > 0$ *and* $s_2 > 0$ **do**
    if $MemSize(\hat{\pi}, s_1) < MemSize(\bar{\pi}, s_2)$ **then**
        $\lfloor$ return $\hat{\pi} < \bar{\pi}$;
    if $MemSize(\hat{\pi}, s_1) > MemSize(\bar{\pi}, s_2)$ **then**
        $\lfloor$ return $\bar{\pi} < \hat{\pi}$;
    $s_1 \leftarrow s_1 - 1; s_2 \leftarrow s_2 - 1$;;
if $s_1 == 0$ **then**
    $\lfloor$ return $\hat{\pi} < \bar{\pi}$;;
else
    $\lfloor$ return $\hat{\pi} > \bar{\pi}$;;

---

features that comprise $\hat{\pi}$ in increasing order of memory sizes, and then returns the memory size of the x'th feature from this ordering.

Model Selection: The model selection for TOMMA happens in a similar fashion to that of TOMMA(S) except for a few subtle differences. Initially $\hat{\pi}_{best}$ is assigned to the model appearing first in the sorted sequence, i.e., the completely stationary model. Assume at some planning phase, $\hat{\pi}_{best} = \hat{\pi}$. Then in the next planning phase, TOMMA *compares $\hat{\pi}$ with all possible models that include all features from $\hat{\pi}$ plus one additional feature*, to check whether there is any incremental model that is more predictive than $\hat{\pi}$. If it finds one, say $\bar{\pi}$, it rejects $\hat{\pi}$ and switches to the next model in its model sequence. The comparison between every pair of such models is performed in the same manner as presented in Equation 6.2.

It is important to note that it does not directly switch to $\bar{\pi}$ as the next model to check because all we can infer is that $\bar{\pi}$ is more predictive than $\hat{\pi}$. But $\bar{\pi}$ can still be sub-optimal, and there may be other models following $\hat{\pi}$ and preceding $\bar{\pi}$ in the sorted model sequence, which are better candidates for π_o and are easier to learn. We can only find them by incrementally searching through the model sequence.

Action Selection: The T-step action selection policy for TOMMA is also very similar to that of TOMMA(S) except for one significant difference. Assume $\mathcal{F} = \{f_1, f_2, f_3, f_4\}$. Assume at some planning phase, $\hat{\pi}_{best} = \hat{\pi}$ where $\hat{\pi}$ is based on features f_1 and f_3. For that phase, TOMMA follows RMAX assuming the state space is a combination of all individual feature spaces comprised of all features from $\hat{\pi}$ plus an additional feature. In the above example, this boils down to the state space being a combination of feature spaces $\{f_1, f_2, f_3\}$ and $\{f_1, f_3, f_4\}$. Thus, for all states which have an unvisited entry (an instantiation of the feature set $\{f_1, f_2, f_3\}$ or $\{f_1, f_3, f_4\}$, which have not been visited m times), it provides the exploration bonus. For all other states, it assumes that the transition and reward functions are determined by $\hat{\pi}$. In spirit similar to TOMMA(S), it then strives to explore all states pertaining to this augmented state space, m times, for evidence suggesting that $\hat{\pi}$ is insufficient for modeling π_o. If it cannot find one, RMAX converges to exploiting based on $\hat{\pi}$.

Like TOMMA(S) and unlike MLEM, TOMMA too may get stuck at a local optimum by converging to a sub-optimal model because of insufficient exploration. As suggested earlier, we consider this to be an acceptable tradeoff.

However, once converged to the correct model $\hat{\pi}_K$, both TOMMA(S) and TOMMA stick to it with a very high probability from then onwards. In fact it is not hard to show that the false negative rate of the model selection component for TOMMA(S) and TOMMA are δ and $n\delta$ respectively. The following paragraph explains why.

Consider first the case of TOMMA(S). All we need to compute is the probability with which TOMMA(S) rejects $\hat{\pi}_K$, once it has rejected all smaller sized models. Assume the worst case that this occurs only after all of the possible $\mathbf{b_K}$'s, and $\mathbf{b_{K+1}}$'s get visited m times. From Equation 6.2, TOMMA(S) can only reject $\hat{\pi}_K$ with error probability at most δ. Suppose it does not reject $\hat{\pi}_K$. Then it should not reject $\hat{\pi}_K$ in the future either, since the estimates of Δ_K and σ_K never change from then onwards (since all of the possible $\mathbf{b_K}$'s, and $\mathbf{b_{K+1}}$'s have already been visited m times, and there can be no future updates to models $\hat{\pi}_K$ and $\hat{\pi}_{K+1}$). Thus the error probability of rejecting $\hat{\pi}_K$ is strictly upper-bounded by δ. The proof for TOMMA follows in a similar fashion. The error probability in this case is $n\delta$, because we have to sum up the error from all model comparisons performed on the correct model, which is upper-bounded by n (size of $\mathcal{F}$).

This concludes our description of the algorithms, and their properties. We now present our first domain for empirical analysis.

6.2 The Surveillance Game

To empirically validate our algorithms, we introduce a challenging new domain — *The Surveillance Game*. The game is motivated by the multi-robot patrol problem, a well studied problem in the robotics community, e.g. [1, 53]. In the general version of the problem, a team of robots is required to repeatedly visit some target area (e.g., perimeter, 2-D environment) in order to maximize its chance of detecting certain adversaries which are trying to penetrate through the patrol path. Although the problem has received considerable attention in recent years, past research tends to seek fixed patrol paths that do not adapt to adversary behavior. To the best of our knowledge, there is no prior work on deploying learning agents for surveillance.

6.2.1 Game Specifics

In the Surveillance Game (see Figure 6.1), the perimeter is divided into P discrete segments. There are k robots monitoring the perimeter, denoted by $R_i, 0 \leq i < k$. Each robot is in charge of a fraction of the perimeter of size $\frac{P}{k}$, with one-segment overlaps at the boundaries. Thus R_0 is in charge of perimeter segments 0 to $\lceil \frac{P}{k} \rceil$, R_1 is in charge of perimeter segments $\lceil \frac{P}{k} \rceil$ to $2\lceil \frac{P}{k} \rceil$, and so forth. There are k intruders denoted by $I_i, 0 \leq i < k$, each attempting to penetrate through one of the boundary segments (henceforth known as the *penetration segments*). Specifically, I_0 tries to penetrate through segment 0, I_1 through segment $\lceil \frac{P}{k} \rceil$, and so on. So $\forall 0 \leq i < k-1$, R_i and R_{i+1} share the job of preventing I_{i+1} from penetrating through penetration segment $\lceil (i+1)\frac{P}{k} \rceil$, while R_{k-1} and R_0 jointly try to prevent I_0 from penetrating through penetration segment 0. Apart from preventing the intruders from penetrating, the robots also have an additional task of periodically patrolling their part of the perimeter, with the idea that they have some other duty to perform along the way, which is not mission critical, but that is best done frequently (such as cleaning).

We assume the presence of a centralized controller that has full visibility, and controls the robots. We decide the action selection on every time step for the controller, not the intruders. The robots can move to an adjacent segment in 1 time step. Each penetration takes $\tau > 0$ time steps to complete. To keep the problem tractable, we assume that there are just two actions available to each robot: whenever a robot reaches a penetration segment, it can either take a *short lap*, or a *long lap*. A short lap is when the robot leaves the penetration segment, and comes back just in time to catch a penetration (if one happened), i.e., it moves $\tau/2$ segments away and then returns so that it's guaranteed to catch any intruder that started to penetrate while it was gone. A long lap is when the robot traverses its complete range of segments, e.g., R_0 traversing from penetration segment 0 to penetration segment $\lceil \frac{P}{k} \rceil$, or vice versa. If a robot reaches a penetration segment while a penetration is taking place, we say that the intruder has been *caught*, with no penalty incurred. On the other hand each successful penetration leads to a penalty of R_{Pen} (or a reward of $-R_{Pen}$). The robots get a positive reward of R_{LO}, and $\xi^Z R_{HI}$ for every short and long lap respectively, with $R_{LO} < R_{HI} < R_{Pen}$. $0 \leq \xi < 1$ is a system level decay constant, and Z is the number of actions elapsed since that robot last took a long lap. This reward structure incentivizes the robots to take long laps periodically.

All of our empirical results focus on the following instance of this domain: $P = 36, k = 3, \tau = 8, R_{LO} = 5, R_{HI} = 100, R_{Pen} = 500$, and $\xi = 0.9$ (Figure 6.1). Henceforth, we always allude to this domain instance.

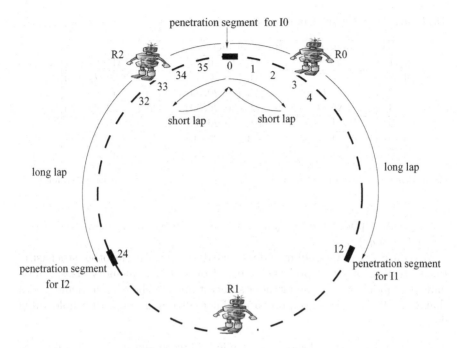

Fig. 6.1 The Surveillance Game

6.2.2 *Intruder Behavior*

Clearly, the intruders' behavior is germane to the problem as based on it the robots respond. We assume that the intruders know everything about the domain except the reward structure and the policy of the robots which are internal to the robots. They also know that the robots can only take short, or long laps. Whether an intruder penetrates on a particular time step depends on its own estimate of the probability with which it might get caught, a.k.a. *probability of penetration detection (ppd)*. We next explain the crucial concept of ppd with an example.

An intruder's estimate of ppd is based on its own estimate of the robots' policy, i.e., how do the two robots guarding the penetration segment balance between taking short and long laps. In order to do that it maintains different estimates of the robots policy based on different memory sizes. Assume I_0 is computing the ppd for a memory size L. This means I_0 keeps track of the ratio of the number of times R_0 and R_2 took short laps in their past L actions. Denote these values as x_{R_0}, and x_{R_2} respectively. To I_0, x_{R_0} is thus an estimate of R_0's policy. Assume a scenario where R_0 and R_2 are at segments 3 and 33 respectively, and are moving away from segment 0 (Figure 6.1). If I_0 decides to penetrate at that time step, the probability with which it believes it is going to get caught by either R_0 or R_2 is $x_{R_0} + (1 - x_{R_0})x_{R_2}$. Thus the

Table 6.1 Different cases of intrusion considered

Case	Population	ppd from Memory sizes	η
1	All S	All $\{10, 20, 30\}$	All .4
2	All S	All $\{10, 20, 30\}$	$I_0, I_1 = .2, I_2 = .4$
3	All D	All $\{10, 20, 30\}$	All .4
4	All D	All $\{10, 20, 30\}$	$I_0, I_1 = .2, I_2 = .4$
5	$I_0, I_1 = $ D, $I_2 = $ S	All $\{10, 20, 30\}$	All .4
6	$I_0, I_1 = $ D, $I_2 = $ S	All $\{10, 20, 30\}$	$I_0, I_1 = .2, I_2 = .4$
7*	All S	All $\{5, 15, 25\}$	All .4

(D stands for a deterministic intruder, and S stands for a stochastic intruder.)

corresponding ppd value for I_0 at that time step based on memory size L is $x_{R_0} + (1 - x_{R_0})x_{R_2}$.

We consider two types of intruders - one deterministic, the other stochastic. Both maintain a set of ppd values based on different memory sizes. These multiple models reflect the intruder's uncertainty in choosing a memory size that best characterizes the ppd value. They also maintain a threshold value denoted by $0 < \eta < 1$.

The *deterministic type* decides to intrude if all its maintained ppd values are $< \eta$, otherwise not.

The *stochastic type* decides to intrude with probability $1 - \eta$, if the minimum ppd value amongst all its maintained ppd values is $< \eta$, otherwise not.

The different scenarios of intrusion consider in our experiments (by varying each of the relevant parameters) is summarized in Table 6.1. For example in Case 1, all the intruders are of stochastic type using ppd values from memory sizes $\{10, 20, 30\}$, and having $\eta = 0.4$.

Note, all the intruders considered are memory-bounded and hence Markovian by definition. However, their memory size can be as big as 30 (Cases 1-6 from Table 6.1). Treating the past 30 joint-actions as features renders the problem intractable. To tackle that, we model the intruders using ppd estimates as features. This gives us a compact way of representing their policy and renders the learning problem tractable. As long we have the correct ppd estimates, ones which are computed from memory sizes 10, 20 and 30 for Cases 1-6, we have the adequate representative capacity to model these intruders. However, maintaining such concise statistics as features makes the underlying setting non-Markovian. This is because the values of these features at time step $t + 1$ is not determined by their collective values and the joint action taken, from time t. In a way we have applied a function approximator [70] over the real Markovian feature space and rendered the problem non-Markovian. Thus our goal is not to converge to an optimal policy since in many scenarios it may not be possible to compute one, but to show that both TOMMA(S) and TOMMA compute decent policies which lead to a high return. In Section 6.2.3, we discuss how we deal with the non-Markovian nature of the underlying setting while computing an action selection policy for our algorithms.

Having introduced the game specifics and the different scenarios of intruder behavior, we next move on to our empirical results.

6.2.3 Results for the Surveillance Game

The most natural comparison point for TOMMA is against current approaches from the literature that tackle memory-bounded agents, such as PCM(A) [58] and CMLES (introduced in Chapter 4). However none of them scale to large memory sizes: in Cases 1-6 from Table 6.1, these approaches would need to explore all joint histories of size 30 to compute the optimal policy, and thus would be prohibitively sample inefficient. Instead, we perform the following comparisons. In the first set of experiments, we compare different variations of TOMMA and TOMMA(S) (with different feature sets as input) to see how they fair under different input settings. While in the second set of experiments, we compare how MLeM(S) and MLeM fair when implemented with the same set of adjustments as suggested in Section 5.6.

6.2.3.1 Experiments Involving TOMMA(S) and TOMMA

Our first set of results focuses on Cases 1-6 from Table 6.1. The different algorithms considered for the controller for these cases are:

1. TOMMA(S) with feature set $\mathcal{F}_1$ (denoted by TOMMA(S)-1) comprised of ppds derived from memory sizes $\{10, 20, 30, 40, 50\}$. Note, $\mathcal{F}_1$ satisfies the sequential structure assumption for all Cases 1-6. In each of these cases the ppds for the intruders are derived from memory sizes 10, 20 and 30: the first three features of $\mathcal{F}_1$;
2. TOMMA(S) with feature set $\mathcal{F}_2$ (denoted by TOMMA(S)-2) comprised of ppds derived from memory sizes $\{10, 15, 20, 25, 30, 40\}$. Note, $\mathcal{F}_2$ does not satisfy the sequential structure assumption, but includes all of the relevant features;
3. TOMMA with feature set $\mathcal{F}_2$;
4. KNOWN-MODEL (denoted by K-M) that assumes full prior knowledge of the relevant features;

Since the ppd values are continuous values $\in [0, 1]$, we discretize the feature space for them into 5 intervals of length 0.2. K-M runs RMAX with $m = 35$. In all our experiments, we seed all variations of TOMMA with values: $\delta = 0.2$, $m = 35$ and $\xi = 0.9$. Our results for cases which involve stochasticity have been averaged over 40 runs (all cases except 3 and 4 from Table 6.1).

Under normal circumstances while dealing with a Markovian agent, the T-step action selection policy on any planning iteration is just the stationary

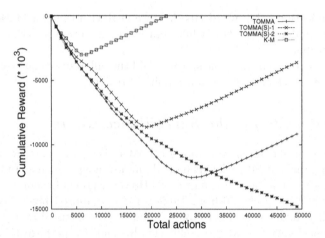

Fig. 6.2 Cumulative reward plot for Case 1

RMAX policy for the underlying MDP executed for T steps (computed using
Value Iteration). However, since we are in a non-Markovian setting, we cannot
solve for a stationary RMAX policy. We counteract this by making a necessary
adjustment to our action selection component. Our planning iteration lasts
for just one time step. However on each time step, we compute a 10-step
episodic RMAX policy and execute the action prescribed by it for the current
time step. This involves creating a lookahead tree of size 10 and choosing the
best action for the current time step based on it.

Our objective is to check how well the different variations of TOMMA and
TOMMA(S) compete with K-M. Note that the final policy computed by K-M
is the best policy that we can expect to compute given the different parameter
values we use for our algorithms. Hence with a slight abuse of terminology,
we refer to the final policy computed by K-M as the *optimal policy* - the one
we strive to achieve.

Figures 6.2 to 6.7 show the cumulative reward plots of all of our algorithms
for Cases 1- 6. Each of these plots show the convergence to the optimal policy
for the different algorithms (convergence is indicated when the upward slope
becomes constant). As expected, all TOMMA variants do much worse than
K-M.

Also as expected TOMMA(S)-1 dominates TOMMA, while TOMMA dominates TOMMA(S)-2, with respect to the convergence to the optimal policy.
TOMMA(S)-1 benefits from having prior knowledge of a feature set ($\mathcal{F}_1$) where
the features satisfy the sequential structure assumption. This explains why it
performs the best. TOMMA(S)-2 also assumes that it has a sequential feature
set, however is provided with an incorrect one ($\mathcal{F}_2$). Hence it has to learn a
much bigger model than TOMMA(S)-1, and does the worst. TOMMA is better
suited for arbitrary feature sets and hence does better than TOMMA(S)-
2. Note in all of these cases, both TOMMA(S)-1 and TOMMA consistently

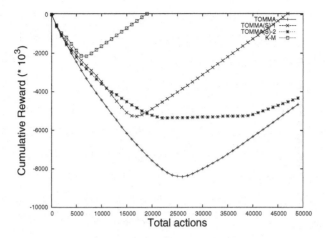

Fig. 6.3 Cumulative reward plot for Case 2

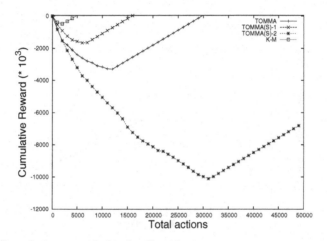

Fig. 6.4 Cumulative reward plot for Case 3

converge to the optimal policy (indicated by almost similar slopes to that of
K-M in all of the Figures 6.2 to 6.7).

Figures 6.8 and 6.9 show all the different models that TOMMA(S) and
TOMMA respectively reject in Case 1, before they converge to the correct
model comprised of ppd values from memory sizes $\{10, 20, 30\}$. The Y-axis
of the two plots shows the different models that TOMMA(S) and TOMMA
skip before they converge to the correct model. A label of [10,20] on the
Y axis refers to a model that is based on ppds derived from memory sizes
10 and 20. Note in the case of TOMMA(S)-1 (Figure 6.8), the sequence of
different models are computed by incrementally selecting features from $\mathcal{F}_1$,
whereas in the case of TOMMA (Figure 6.9) the sequence of different models

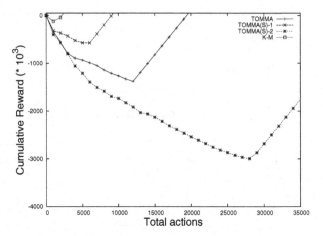

Fig. 6.5 Cumulative reward plot for Case 4

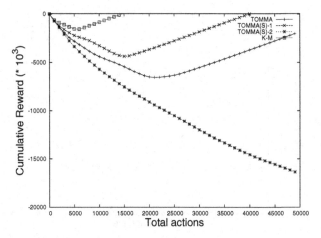

Fig. 6.6 Cumulative reward plot for Case 5

are computed by sorting all possible models based on the sort comparator introduced in Algorithm 10. In the latter case, models which are easier to learn (i.e., ones which are based on smaller memory sizes) take precedence. Since the model selection for all the other cases (Cases 2 - 6 from Table 6.1) are qualitatively the same, we omit their plots.

In all of these experiments, the difference between the converged values (both number of time steps taken to converge to the correct model and the optimal policy) between every pair of algorithms, for all the cases, is statistically significant (by T-test, p-value < 0.05). Also very importantly, both TOMMA and the two versions of TOMMA(S), once converged to the correct model, never switch to a bigger model. This stability provides empirical

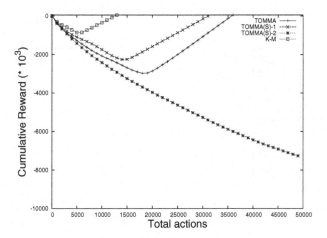

Fig. 6.7 Cumulative reward plot for Case 6

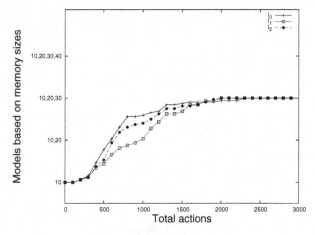

Fig. 6.8 Model selection for TOMMA(S)-1 for Case 1. The I_0 plot shows the model selection for intruder I_0, and so forth.

evidence in support of our claim that the false negative rate of the model selection component of both TOMMA(S) and TOMMA is low.

Our next results concern Case 7 from Table 6.1. In this case, we employ TOMMA, and TOMMA(S) with a feature set $\mathcal{F} = \{10, 20, 30, 40, 50\}$. Note, the feature set does not have the relevant features for the intruders strategies' since the latter is determined by ppds derived from memory sizes 5, 15 and 25. Again, K-M assumes full prior knowledge of the relevant features. Figure 6.10 shows the convergence to the final policy for the different algorithms. Note, neither of our algorithms converges to the optimal policy, the one K-M converges to (different slopes). However, each learns a decent sub-optimal

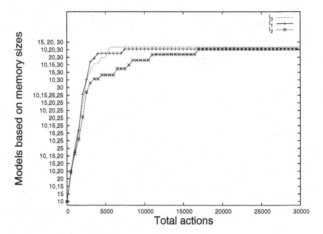

Fig. 6.9 Model selection for TOMMA for Case 1. The I_0 plot shows the model selection for intruder I_0, and so forth.

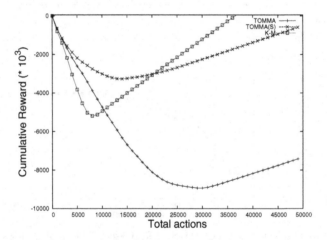

Fig. 6.10 Cumulative reward plot for Case 7

model of the intruders to ensure a competitive return. This result shows that even if $\mathcal{F}$ does not contain the relevant features, but has reasonably good ones, our algorithms can still converge to competitive final policies.

6.2.3.2 Experiments Involving MLeM(S) and MLeM

In our second set of experiments, we tested with MLeM(S) and MLeM as the algorithm used by the centralized controller. We make the same set of empirical adjustments for these algorithms as we did in Chapter 5. We seed these algorithms with an m, δ and $\mathcal{F}$. m plays the role of m_k and is the same

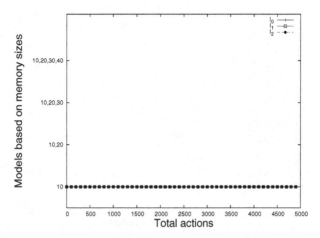

Fig. 6.11 Model selection for MLEM(S) for Case 1. The I_0 plot shows the model selection for intruder I_0, and so forth.

for models of all sizes. δ is required to compute the value of σ_k (Equation 6.3). All our results are reported for $m = 35$ and $\delta = 0.2$. For MLEM(S), $\mathcal{F}$ is comprised of ppds derived from memory sizes $\{10, 20, 30, 40, 50\}$ (the set $\mathcal{F}_1$ from our previous set of experiments), whereas for MLEM, $\mathcal{F}$ comprised of ppds derived from memory sizes $\{10, 15, 20, 25, 30, 40\}$ (the set $\mathcal{F}_2$ from our previous set of experiments). Clearly in case of MLEM(S), $\mathcal{F}$ satisfies the sequential structure assumption. Also alike Chapter 5, we ran MLEM(S) and MLEM without any exploratory iterations, relying solely on the exploration from the greedy iterations.

Like our implementation of MLEM, the planning iteration in this case also lasts for just one time step. On each time step, we compute a 10-step episodic RMAX policy and execute the action prescribed by it for the current time step. This involves creating a lookahead tree of size 10 and choosing the best action for the current time step based on it. As suggested earlier, this can be seen as a counter measure to dealing with the underlying non-Markovian setting.

Unfortunately in all cases from Table 6.1, both MLEM(S) and MLEM never converge to the correct model. In all of these cases, they converge to exploiting based on a shorter incorrect model and achieve a sub-optimal return. The only exceptions are Cases 3 and 4, where despite exploiting based on a shorted model, they learn to follow the optimal policy. Figure 6.11 shows that for Case 1, MLEM(S) converges to exploiting based on a model based on ppd derived from memory size 10. Whereas Figure 6.12 shows that for Case 1, MLEM though learns a decent model based on ppds derived from bigger memory sizes, yet in the end fails to learn the correct model. Since the other model selection plots for Cases 2 - 6 are qualitatively similar to that of Figures 6.11 and 6.12, we omit them.

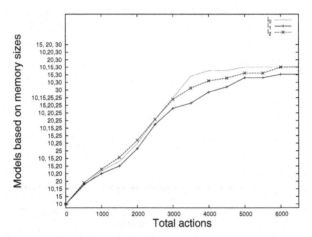

Fig. 6.12 Model selection for MLEM for Case 1. The I_0 plot shows the model selection for intruder I_0, and so forth.

Now the reader might wonder why MLEM(S) and MLEM both failed to learn the correct model, whereas TOMMA(S) and TOMMA succeeded. We believe the reason is that in case of the latter, the algorithms are deliberately designed to explore the next bigger model(s) in the model sequence and only converge to exploiting when the next bigger model(s) do not contribute to the predictiveness. In that fashion the exploration bias helps it to hop over sub-optimal models and try bigger models which appear to be more predictive. On the contrary our approximate empirical implementation of MLEM(S) and MLEM, relies solely on exploring based on the current model. Though this still might lead to decent exploration and convergence to locally good models, its exploratory bias is not as sophisticated as that of TOMMA(S) and TOMMA.

We now move on to present results from our second domain of interest.

6.3   The Ticket Checking Domain

The Ticket Checking domain is inspired by a real life problem of catching passengers who do not buy a ticket (or *evaders*) while traveling on trains. For example in urban transit systems such as the Los Angeles Metro Rail system, passengers are legally required to buy a ticket before boarding a train, but there are no checkpoints prior to boarding which physically deny evaders from boarding a train. Instead patrol officers are deployed in the transit system to check for evaders. A key research question is how to intelligently schedule these patrol officers so that the chances of catching evaders is maximized. Currently the state-of-the-art approach to solving the problem is a stationary Stackelberg policy which assumes that the evaders a-priori know the strategy of the patrol officers and best respond to it [42]. In this section we deal with possibly more realistic evaders who decide their current step action (whether to buy a ticket or not) based on their observations from the past few days. Our results show that TOMMA(S) and TOMMA significantly outperform the Stackelberg policy in such scenarios. We begin by introducing the domain, and then move on to present our results.

6.3.1 Domain Specifics

Our implementation of the domain is similar to the one specified in [42]. The *transit system* is characterized by the number of *stations*, denoted by *Stations*, and the number of *time units* denoted by *Time*. The *train system* consists of a single line, that is all the trains travel in the same direction visiting the same set of stations at specific time points. For simplicity, we assume that the time taken by a train to travel between any two stations is always the same. So we can model time as slotted, focussing only on time points at which some train arrives at a particular station.

For example consider the transit system from Figure 6.13. There are three 3 stations in the system, denoted by A, B and C. Each train starts at station A, visits station B at 1 time point and finally terminates at station C at another time point. Let the time point when a train leaves station A be t. Then the path of that train can be defined as $At - B(t+1) - C(t+2)$. We identify each train by its unique train path. So the 6 possible trains in this transit system are $A1 - B2 - C3, A2 - B3 - C4, A3 - B4 - C5, A4 - B5 - C6$, $A5 - B6 - C7$ and $A6 - B7 - C8$.

We assume that the total number of passengers $\mathcal{P}$ using the system and their distribution across the different routes remain the same every day. Each passenger takes his preferred *route* regardless of the patrol strategy. Also each passenger boards a train from a specific station and leaves at a specific station in a train path. We further assume that each passenger takes at least one time unit to exit the station once he leaves the train. This is to allow some time for a patrol officer to check for his ticket given there exists one to perform that check. So the 3 possible passenger routes for the train $A1 - B2 - C3$ are $A1 - B2 - C2, A1 - B2 - C3 - C4$ and $B2 - C3 - C4$. Note that in each of these routes, each passenger stays at the exiting station for at least 1 time unit. In our example from Figure 6.13, there are $3 \times 6 = 18$ such passenger routes.

There are a fixed number γ of deployable *patrol officers*, each of whom can be scheduled for at most ξ time units. There are two types of atomic patrol actions: *on-train inspection* where an officer checks for tickets while traveling on a train from one station to the other, or *in-station inspection* where an officer checks for tickets of off-boarding passengers at a particular station, each lasting for 1 time unit. Thus a feasible *strategy for any patrol officer* can be any sequence of such atomic actions of size ξ. For example

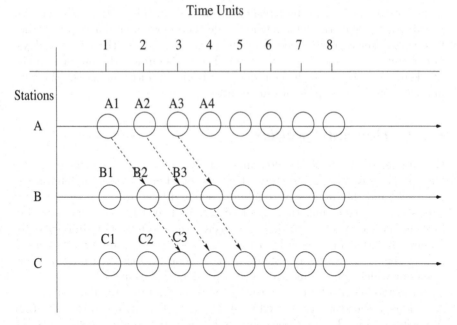

Fig. 6.13 The transit system with 3 stations and 8 time units

if $\xi = 3$, a possible strategy for a patrol officer for the transit system from
Figure 6.13 can be $A1 - B2 - C3 - C4$, which comprises 2 on-train inspections,
namely $A1 - B2$ and $B2 - C3$, and 1 in-station inspection $C3 - C4$. Similarly
$C3 - C4 - C5 - C6$ is another such possible patrol strategy which comprises
3 in-station inspections, namely $C3 - C4$, $C4 - C5$ and $C5 - C6$.

However, just because an evader travels in a train with a patrol officer
does not mean that he is going to get caught. It might be that the evader
is standing at the very end of the train and there may be not be enough
time for the patrol officer to reach all the way to the end of the train to
catch him. So given a patrol strategy P and a rider route Z, the inspection
probability of an evader on that route for that patrol strategy is given by
$min\{1, \sum_{e \in P \wedge Z} f\}$. $f \in [0, 1]$ is a system defined fixed probability with which
an evader gets caught for an atomic patrol action on his route. We justify the
inspection probability as follows. We assume that for on-train inspections,
the riders are inspected one at a time starting from the start of the train.
The fraction of the train that is inspected depends on the duration of the
on-train inspections. Given sufficient number of such on-train inspections,
the patrol works his way through the entire train and eventually catches an
evader at the end of the train. Similarly for in-station inspections, a patrol
officer can be assigned to any random compartment of the train and can only
inspect a fraction f of the total volume of passengers. That explains why the

inspection probability adds up. The total inspection probability is the sum of the inspection probabilities from all patrol strategies, each pertaining to a patrol officer.

We assume that each passenger is *risk neutral*. That is each passenger makes a binary decision of buying a ticket, or not, based on his expected cost from performing the two actions on that particular day. For example, let the price of a ticket be $Fare$ and the fine for fare evasion when caught be $Fine$. Also assume that the passenger's internal estimate of the probability of getting detected on that particular day (explained in the next paragraph) for not buying a ticket be p. Then a risk neutral passenger prefers not to buy a ticket as long as the following inequality holds, $p \times Fine < Fare, or, p < \frac{Fare}{Fine}$.

Now all that remains to be explained is how each passenger estimates his probability of getting detected on a particular day for not buying a ticket, or *pod*. We assume that the passengers get to see the patrol officers' strategies on each day [1]. Each passenger looks back at a fixed number L of past samples of patrol officers' strategies (from last L days) to compute an estimate of the latter. For example let the passenger route be $A1 - B2 - B3$ and $L = 3$. Assume that in the past 3 days, there has been only one $A1 - B2$ on-train inspection and no $B2 - B3$ in-station inspections. So his estimate is that the $A1 - B2$ on-train inspection and the $B2 - B3$ in-station inspection happens each day with a probability of $\frac{1}{3}$ and 0 respectively. Hence the corresponding *pod* estimate is, $p = \frac{1}{3} \times f + 0 \times f = \frac{1}{3}f$.

This concludes our specification of the domain. We next move on to present our results for this domain.

6.3.2 Results for the Ticket Checking Domain

Our results are for a specific instance of the domain with $Stations = 3$, $Time = 8$, $\mathcal{P} = 1000$, $Fine = 25$, $Fare = 10$, $f = 0.3$, $\gamma = 1$ (meaning one patrol officer) and $\xi = 5$. Figure 6.13 shows the transit system. The population of 1000 passengers is distributed randomly to each of the 18 possible routes on each run of our simulation. For each route, we have a mix of passengers who compute their *pod* estimates by using $L \in \{2, 3, 4\}$. Thus for each passenger route the fraction of the population that evades is determined by the values of 3 unknown features, namely the pod estimates of that route from memory sizes 2, 3 and 4.

Our results focus on using TOMMA(S) and TOMMA for determining the patrol strategy on each day. Akin to the setting in [42], we assume prior knowledge of the total number of passengers in each route for each simulation run. TOMMA (and also TOMMA(S)) maintains different models for each different passenger route. Neither of these algorithms have any prior knowledge of the exact features. However they assume that the features are drawn from a set

---

[1] Assuming that the word goes around each day on how the patrolling was performed on that particular day.

$\mathcal{F}$ which comprises pod estimates from values of $L \in \{1, 2, 3, 4, 5, 6\}$. A *model for a route* is a mapping from a set of pod estimates computed from different values of L to the fraction of evaders in the population of that route. Since the pod values are continuous values $\in [0, 1]$, we discretize the feature space for them into 5 intervals of length 0.2. In all of our experiments, we seed all variations of TOMMA with values: $\delta = 0.2$ and $m = 1000$.

Note that the pod estimates are non-Markovian features as well (just like ppds). Thus we make a necessary adjustment to our action selection component. Our planning iteration lasts for just 1 time step. On each time step, we compute a 2-step episodic RMAX policy and execute the action prescribed by it for the current time step. This involves creating a lookahead tree of size 2 and choosing the best action on the current time step based on it.

Our first benchmark for comparison is the Known-Model (or K-M) version which assumes prior knowledge of all of the relevant features for each passenger route and runs RMAX with $m = 1000$. Note that the final policy computed by K-M is the best policy that we can expect to compute given the different parameter values we use for our algorithms. Hence with a slight abuse of terminology, we refer to the final policy computed by K-M as the *optimal policy* - the one we strive to achieve.

Our second benchmark for comparison is a random strategy, called Random, that randomly allocates a patrol strategy for our patrol officer from amongst the possible patrol strategies.

Finally, our third benchmark for comparison is the Stackelberg solution from [42], called Stackelberg. Note, the total expected payoff (money collected on each day from both fines and fares) can be decomposed into individual expected payoffs from bilateral interactions between each individual passenger and the patrol officer. This implies that the game is payoff equivalent to a Bayesian Stackelberg game between one patrol officer (leader) and one passenger (follower) whose type is determined by a prior probability distribution on the different passenger routes. The probability that the passenger is from passenger route r is proportional to the ratio of the passenger volume on that route to $\mathcal{P}$. Since the game is a zero-sum game, the Stackelberg solution is the max-min solution of this Bayesian Stackelberg game. For our domain, we can easily compute it using a Linear Program.

All of our results have been averaged over 40 runs. As mentioned earlier, in each such run we randomly assign the route for each passenger to one of the possible 18 routes. Also for each passenger, the distribution from which they are assigned one of the values from $L \in \{2, 3, 4\}$ (needed to compute the pod) is chosen randomly on each such run.

Figure 6.14 shows the cumulative reward plots of all of our algorithms. As expected, both the TOMMA variants do much worse than K-M. Also as expected the two variants of TOMMA dominate Stackelberg and Random. It is interesting to see that both the TOMMA variants do equally well. The difference in the cumulative reward obtained after 5000 simulated days between the two variants of TOMMA is statistically insignificant by a T-test.

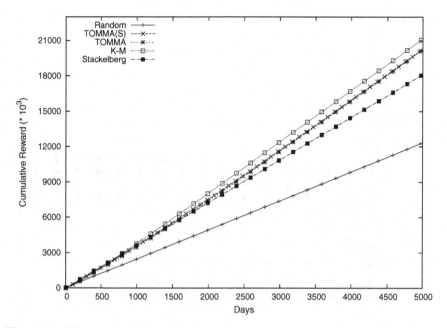

Fig. 6.14 The cumulative reward plot

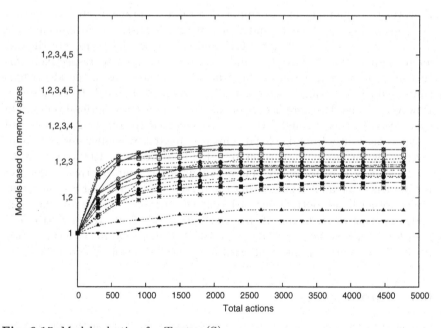

Fig. 6.15 Model selection for TOMMA(S)

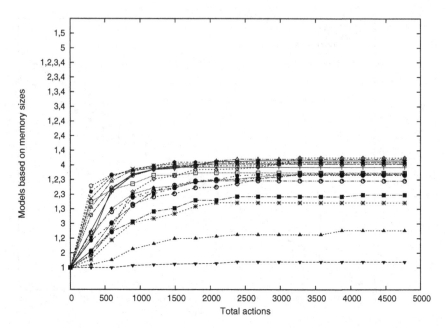

Fig. 6.16 Model selection for TOMMA

However the difference in the cumulative reward obtained after 5000 simulated days between TOMMA and Stackelberg is statistically significant by a T-test (p-value < 0.05). Figures 6.15 and 6.16 show the average of the different models that TOMMA(S) and TOMMA converged to respectively, for the different passenger routes. Note in both of these cases, the algorithms converged to exploiting based on smaller sub-optimal models for each route. However, the models were decent enough to generate the optimal policy. On average both TOMMA and TOMMA(S) converged to following the optimal policy consistently from day 3500. This is indicated by similar upward slopes of these algorithms in comparison to that of K-M from day 3500 and onwards.

Stackelberg is a stationary policy that does not adapt to the behavior of the evaders. Instead it assumes that the passengers always best respond to it. On the other hand, both of our variations of TOMMA leverage from learning and exploiting the behavior of the evaders to good effect. To the best of our knowledge, ours is the first attempt to model the behavior of evaders as being feature based and to implement a learning based solution to catch them.

6.4 Summary

Our goal in this chapter was to propose simpler variants of the MLEM(S) and MLEM algorithms introduced in Chapter 5. The specification of both of these algorithms were motivated from a theoretical standpoint and empirical implementation required certain adjustments. The algorithms introduced in this chapter, namely TOMMA(S) and TOMMA, are easier to implement than MLEM(S) and MLEM respectively, and can be implemented straight from their specification without any empirical adjustments.

Just like MLEM (and hence MLEM(S)), our setting of interest in this chapter is a repeated game between an agent under our control and another Markovian agent. We assume prior knowledge of a set of features $\mathcal{F}$ some of which are assumed to characterize the unknown policy of the Markovian agent. Being unaware of the exact feature set that determines the Markovian agent's policy, both TOMMA(S) and TOMMA maintain different models of the latter based on different combinations of features from $\mathcal{F}$. Then in online repeated play, the main objective of these algorithms is to figure out which amongst these models successfully determines the policy of the Markovian agent and consequently plan its actions based on that. Both TOMMA(S) and TOMMA successfully addresses both of the above problems by leveraging insights from the RMAX algorithm. They are fully implemented with results from two challenging surveillance based domains demonstrating their effectiveness.

This concludes our discussion on modeling other agents in a repeated game setting. In Chapters 3 to 6, we presented different algorithms that provide different sets of guarantees depending on the different assumptions they make of the agents they are interacting with. In the next chapter we take a departure from the repeated game setting, the recurrent theme in the book so far, and focus on a specific instance of a single agent RL problem, called structure learning for factored MDPs. Here we show how our solution to modeling a Markovian agent as addressed in Chapter 5, extends to solving this RL problem with new sample complexity bounds which are competitive and compares well with the state-of-the-art approach in representative domains.

Chapter 7
Structure Learning in Factored MDPs

In Chapter 5 we proposed a couple of subroutines, namely MLeM(S) and MLeM, that are tailored to model and exploit the unknown policy of a Markovian agent. Both of these subroutines assume prior knowledge of the possible set of features upon which a Markovian agent may base its policy, but not the exact set. For both of these subroutine, we pose the problem of modeling the unknown policy of a Markovian agent as learning the unknown feature space and transition function of an induced MDP (induced by the Markovian agent's policy). Now it so happens that both of these subroutines can be easily extended to solve a related challenging problem in Reinforcement Learning (RL), namely structure learning in factored MDPs. The purpose of this chapter is to lay down the direct correlations between the two aforementioned problems and show how the MLeM(S) and MLeM subroutines extend to solve the structure learning problem for factored MDPs.

Representing a MDP [70] with a large state space is challenging due to the curse of dimensionality. One popular way of doing so is factoring the state space into discrete factors (a.k.a features), and using formalisms such as the Dynamic Bayesian Network (DBN) [36] to succinctly represent the state transition dynamics. A DBN representation for the factored transition function can capture the fact that the transition dynamics of a factor is often dependent on only a subset of other factors, called *parent factors* of that factor (see Figure 7.1). The size of the biggest set of parent factors is the *in-degree* of the DBN. Such an MDP is called a factored state MDP, or simply factored MDP (FMDP). We refer the reader to Definition 5 from Chapter 2 to re-familiarize oneself with the formal concept of FMDPs. From a high level, this chapter addresses the problem of learning the unknown factored transition function of each individual factor and planning based on it, by efficient exploration and exploitation. In the RL parlance, the complete problem is often called structure learning in FMDPs [4, 30].

There are three important parallels between the problem of modeling a Markovian agent and structure learning in a FMDP. First, in both cases the underlying framework is a MDP (an AIM in case of the former, see

D. Chakraborty, *Sample Efficient Multiagent Learning in the Presence of Markovian Agents*, Studies in Computational Intelligence 523, DOI: 10.1007/978-3-319-02606-0_7, © Springer International Publishing Switzerland 2014

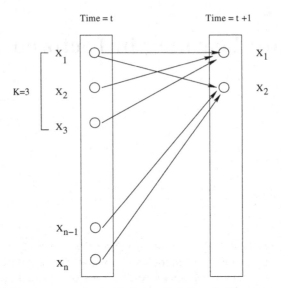

Fig. 7.1 A DBN representation with in-degree $K = 3$. The next step values for factors X_1 and X_2 for a particular action depend only on the current step values of factor sets $\{X_1, X_2, X_3\}$ and $\{X_1, X_{n-1}, X_{n-3}\}$ respectively.

Definition 17 from Chapter 2). Second, in both cases the structure of the transition function of the MDP is unknown. In case of a Markovian agent, we are unsure which features determine the former's policy. In case of a FMDP, we are unsure which parent factors determine the transition of a factor. Third, in both cases we strive to achieve an actual return close to U^* in provably efficient sample complexity. In case of modeling a Markovian agent, U^* is the expected return from following the optimal policy in the induced AIM. In case of structure learning in a FMDP, U^* is the expected return from following the optimal policy in the FMDP. Having observed these parallels for the rest of this chapter we focus solely on the problem of structure learning in FMDPs.

Just like we did in all previous chapters, we assume that the FMDP is a unichain one. As illustrated in Chapter 2, there are a number of key theoretical properties of unichain FMDPs (or more generally unichain MDPs) which simplify our analysis. Foremost amongst them is that while trying to compute an optimal policy for the FMDP, we do not need to worry about different returns originating from different states. The optimal return is a unique value denoted by U^* and can be achieved from any start state by following the optimal policy. For the rest of the chapter whenever we allude to a FMDP, we mean a unichain FMDP.

Some of the earlier work in RL for FMDPs, are extensions of provably sample efficient RL algorithms for general MDPs. Algorithms such as Factored-E^3 [46] and Factored-RMAX [36] achieve near optimal behavior in a polynomial number of samples, but require prior knowledge about the structure of the transition function, in the form of complete DBN structures with unknown conditional probability tables. More recently, there have been approaches proposed that do not assume prior knowledge about the structure of the transition function, and constructs it from experience in the environment [29, 39]. However, these results are mostly empirical successes in particular domains, with no formal analysis of sample complexity.

To the best of our knowledge, SLF-RMAX [4] is the first algorithm to solve the structure learning problem with a formal guarantee on sample complexity. More recently, [30] proposed MET-RMAX, which improves upon SLF-RMAX with a better sample complexity guarantee. However, akin to SLF-RMAX, MET-RMAX requires as input the in-degree (K) of the transition function's DBN structure. In this chapter, we are interested in scenarios where prior knowledge of K is unavailable and K is hard to guess.

Our objective is to propose the first structure learning algorithm for ergodic FMDPs, called *Learn Structure and Exploit with* RMAX (LSE-RMAX), that leverages from our findings in implementing MLEM, to efficiently solve the problem without requiring a prior knowledge of the in-degree K. Instead we make a different assumption that is analogous to Assumption 1 from Chapters 4 and 5, stated next.

Like MLEM, LSE-RMAX operates by planning for T time steps at a time. In each such planning iteration, it uses the best model of each factored component of the transition function, namely each $P_{i,a}$, at hand and plans its actions for the next T time steps based on it. To facilitate the theory behind our claim that LSE-RMAX eventually achieves a return very close to U^*, we assume that the (ϵ, T) pair taken as input always satisfies the following condition:

Assumption 2. *The planning horizon T is sufficiently large and the ϵ sufficiently small to ensure that*

1. *T is the ϵ-return mixing time of the optimal policy for the AIM;*
2. *for any sub-optimal policy π and for any state s of the induced AIM, $U_T^\pi(s) < U^* - 2\epsilon$;*

A pertinent question is whether for any FMDP such an (ϵ, T) pair exists or not. Let $\hat{U}$ be the expected return in the FMDP from the best sub-optimal policy. Lets choose an ϵ smaller than $\frac{U^* - \hat{U}}{3}$. Let T be the maximum of all ϵ-return mixing times from all policies. Clearly this choice of an (ϵ, T) pair satisfies Assumption 2. Hence for any FMDP, there exists an (ϵ, T) pair that satisfies Assumption 2.

Our analysis in this entire chapter assumes that LSE-RMAX is aware of such an (ϵ, T) pair that satisfies Assumption 2. In cases where it is unaware of such an (ϵ, T) pair, the approach presented in Section 4.1.5 which extends MLES to solve this general case, applies here as well. Having noted that we ignore this case in our analysis.

We propose two variants of LSE-RMAX, for two versions of the problem.

1. The first deals with a specific case where factors for each individual $P_{i,a}$ can be arranged in a sequence with the relevant factors (which determine the transition function) preceding the irrelevant ones in the sequence (analogous to the sequential structure learning problem from Section 5.2, see Definition 21).

   For example, the probability of a machine running on a particular time step in a network may depend on the states of its K closest neighbors from the past step. Though this fact is common knowledge, K may be unknown. The sequence of factors is then an incremental ordering of the states of all neighbors based on adjacency, with the immediate neighbors heading the sequence. We call problems of such a nature *sequential structure learning problems for FMDPs*. We believe many real world problems can be mapped to problems of such nature, and hence can leverage from our solution.

2. The second deals with the general case where such an ordering is unknown. Problems of such nature constitute the more general *structure learning problem for FMDPs* (analogous to the general structure learning problem from Section 5.3).

Since the sample complexity analysis for LSE-RMAX is similar to that of MLEM, our main goal in this chapter is to tie the knot between the sample complexity analysis of the two approaches and derive similar style bounds as derived in Chapter 5. Our bounds are competitive and compare well with that of MET-RMAX. We also present empirical results from two popular benchmark domains demonstrating LSE-RMAX's competitiveness.

The remainder of the chapter is organized as follows. Sections 7.1 and 7.2 present LSE-RMAX in detail, Section 7.3 presents our empirical analysis and Section 7.4 summarizes the chapter.

7.1   Sequential Structure Learning in FMDPs

As suggested earlier, we use the same technique as used in solving the sequential structure learning problem for modeling a Markovian agent in solving its counterpart for FMDPs. We refer the reader to Section 5.2 (the discussion on MLEM(S)) for a complete account of how to solve the sequential structure learning problem. Our goal in this section is to tie the knot between our current approach and the one presented in Section 5.2 to derive similar style sample complexity bounds. Analogous to MLEM(S), we call our variant of LSE-RMAX tailored to solve this specific problem as LSE-RMAX(S).

Recall from Definition 5 that in a FMDP, each state consists of n discrete factors. Furthermore the transition function of the FMDP is also factored into $n \times |A|$ components, denoted by $P_{i,a}$'s, $i \leq n, a \in A$, and each $P_{i,a}$ satisfies the conditional independence assumption. Each $P_{i,a}$ has its own structure. We assume that the in-degree of the DBN structure is K: all factors have at most K parent factors.

Trivially, the problem can be solved by assuming that every $P_{i,a}$ is dependent on all n factors, and running RMAX. The downside of this approach is that it requires an order of $\approx \tilde{O}(d^n)$ data samples, where d is the maximum of the number of values each $P_{i,a}$ can assume. Needless to say, the above approach does not scale for large values of n. Our objective is to solve the problem in a sample complexity which is polynomial in d^K. We begin by formally defining the sequential structure learning problem for FMDPs.

Definition 25. *Sequential Structure Learning problem for FMDPs: We assume that for each $P_{i,a}$, we have an ordering of the parent factors and the first K factors in the ordering determine $P_{i,a}$, K being unknown. The learning problem is how to efficiently model each $P_{i,a}$ given this rich factor space representation and achieve a return close to U^* with a high likelihood, in the best possible sample complexity.*

The main objective of LSE-RMAX(S) is to learn a decent approximation of each $P_{i,a}$, such that the policy computed from such a joint model yields a return close to U^*. LSE-RMAX(S) then treats each $P_{i,a}$ in the same fashion as MLEM(S) treats a Markovian agent whose feature space comprises the first K features from a set $\mathcal{F}$. So it maintains $n \times |A|$ number of MLEM(S) instances, one pertaining to each $P_{i,a}$. On any T-step planning iteration, the FIND-MODEL(S) component of each MLEM(S) instance returns a specific model for its corresponding $P_{i,a}$ based on which the action selection module must operate. The only exception is during the exploratory planning iteration, where LSE-RMAX(S) in a principled manner chooses a random model for each $P_{i,a}$ to explore.

For the action selection part, just like MLEM(S), LSE-RMAX(S) interleaves between executing greedy planning iterations and exploratory planning iterations. The RMAX policy computation for every greedy planning iteration happens as follows. Let $\hat{P}_{i,a}$ denote the model chosen by the FIND-MODEL(S) component of the MLEM(S) instance associated with $P_{i,a}$. Then for any state action pair (s, a), if there exists an (i, a) pair such that $\hat{P}_{i,a}(s) = \perp$, it means that LSE-RMAX(S) has no way of estimating $P_{i,a}(s)$ based on $\hat{P}_{i,a}$. To facilitate exploration to such state action pairs, LSE-RMAX(S) gives them the imaginary exploratory bonus. For all other state action pairs, LSE-RMAX(S) performs the conventional Bellman backup.

On a similar note, the RMAX policy computation for every exploratory planning iteration happens as follows. Let the random model chosen for $P_{i,a}$ be $\hat{P}_{i,a}$. Then for any state action pair (s, a), there exists an (i, a) pair such that $\hat{P}_{i,a}(s) = \perp$, LSE-RMAX(S) gives that state action pair the imaginary exploratory bonus. For every other (s, a) pair, it uses the model for $P_{i,a}$ predicted by its corresponding FIND-MODEL(S) which is the best predictive model for the corresponding $P_{i,a}$ based on the data samples seen so far, to perform the Bellman back up.

Then through a sample complexity analysis very similar to that of Lemma 5.2.2, we appear at the sample complexity bound for LSE-RMAX(S).

But there are a few differences. First, compared to a state space size of N_K in Lemma 5.2.2, the current state space is of size $n|A|d^K$. This is because we have $n \times |A|$ number of $P_{i,a}$'s, and each span over d^K values. Second, we have a new value of m_k (as opposed to Equation 5.5). The new value of m_k is computed accounting for the following differences:

- compared to a requirement of an $\frac{\epsilon}{T}$-approx model for π_o in Lemma 5.2.2, here we need to learn an $\frac{\epsilon}{nT}$-approx model for each $P_{i,a}$. The reason is that since the state space is composed of n independent factors, an error of $\frac{\epsilon}{nT}$ in the estimation of transition for each factor, leads to a total error of $\frac{\epsilon}{T}$ in the estimation of transition for the entire state;
- more so we need to learn an $\frac{\epsilon}{nT}$-approx model of each $P_{i,a}$ with an error probability of at most $\frac{\delta}{n|A|}$. This means we have learnt the entire factored transition function with an error probability of at most δ;
- finally, in case of MLEM(S) each model specifies a distribution over $|A|$ actions. Thats why the term $|A|$ in the numerator of the log term in Equation 5.5. Here, each model $\hat{P}_{i,a}$ specifies a distribution over d items;

Accounting for the above mentioned differences, gives us our new value for m_k.

$$m_k = O(\frac{n^4 T^2}{\epsilon^2} log(\frac{n^2 |A| d^{k+1}}{\delta})) \tag{7.1}$$

This brings us to our main theoretical result concerning LSE-RMAX(S), stated next.

Lemma 7.1.1. *For any $0 < \epsilon < 1$ and $0 < \delta < 1$, with a high probability of at least $1 - 4\delta$, LSE-RMAX(S) achieves an actual return $\geq U^* - 5\epsilon$ for the sequential structure learning problem for FMDPs, in a number of time steps given by*

$$O(\frac{n^6 |A| d^K T^3}{\epsilon^7} log(\frac{n^2 |A| d^{K+1}}{\delta}) log^2(\frac{1}{\delta})),$$

a quantity polynomial in $\frac{1}{\epsilon}$, $\frac{1}{\delta}$, n, d^K, $|A|$ and T.

That concludes our discussion of LSE-RMAX(S), our algorithm for solving the sequential structure learning problem for FMDPs. Next we present the full blown LSE-RMAX algorithm that uses the MLEM algorithm from Chapter 5 to solve the general structure learning problem.

7.2   General Structure Learning in FMDPs

In the general version of the problem, the ordering of the factors for each $P_{i,a}$ is arbitrary (based on the best possible guess) and not by relevance. In other words, there might be irrelevant factors preceding the relevant factors in each ordering. Since the technical specification of LSE-RMAX is very similar to that of MLEM, we do not present the former in detail. We refer the reader to Section 5.3 (the discussion on MLEM) for a complete account of how to solve the general structure learning problem. Our goal in this section is to tie the knot between our current approach and the one presented in Section 5.3 to derive similar style sample complexity bounds.

LSE-RMAX treats each $P_{i,a}$ in the same fashion as MLEM treats a Markovian agent whose feature space comprises any K features from a set $\mathcal{F}$. So it maintains $n \times |A|$ number of MLEM instances, one pertaining to each $P_{i,a}$. On any planning iteration, the FIND-MODEL component of each MLEM instance returns a specific model for its corresponding $P_{i,a}$ based on which the action selection module must operate. The only exception is during the exploratory planning iteration.

For the action selection part, just like MLEM, LSE-RMAX interleaves between executing greedy planning iterations and exploratory planning iterations. The RMAX policy computation for every greedy planning iteration happens as follows. Let $\hat{P}_{i,a}$ be the model chosen by the FIND-MODEL component of the MLEM instance associated with $P_{i,a}$. Then for any state action pair (s,a), if there exists an (i,a) pair such that $\hat{P}_{i,a}(s) = \perp$, it means that LSE-RMAX has no way of estimating $P_{i,a}(s)$ based on $\hat{P}_{i,a}$. To facilitate exploration to such state action pairs, LSE-RMAX gives them the imaginary exploratory bonus. For all other state action pairs, LSE-RMAX performs the conventional Bellman backup.

On a similar note, the RMAX policy computation for every exploratory planning iteration happens as follows. LSE-RMAX chooses a k randomly from 0 to n. For all (s,a) pairs that contain a k factor value $\mathbf{b_k}$ in s for which action a has not been taken m'_k times, give them the imaginary exploratory bonus. For every other (s,a) pair, use the model for $P_{i,a}$ predicted by its corresponding FIND-MODEL which is the best predictive model for the corresponding $P_{i,a}$ based on the data samples seen so far, to perform the Bellman back up.

Then through a sample complexity analysis very similar to that of Lemma 5.3.2, we appear at the sample complexity bound for LSE-RMAX. The key differences are as follows.

First, compared to a state space size of $\mathbf{N_{2K}}$ in Lemma 5.3.2, the current state space is of size $\binom{n}{2K}|A|d^{2K}$. This is because our exploration condition from Lemma 5.3.1 requires that for each individual $P_{i,a}$, every $2K$-dimensional factor value be visited a sufficient number of times and there are $\binom{n}{2K}d^{2K}$ number of $2K$-dimensional factor values.

Second, we have a new value for m'_k (as opposed to Equation 5.6). The new value of m'_k is computed accounting for the following differences:

- compared to a value N'_k for the maximum size of the feature space from any k-dimensional feature vector in Lemma 5.3.2, the current value of the same is d^k;
- as observed earlier, compared to a requirement of an $\frac{\epsilon}{T}$-approx model for π_o in Lemma 5.2.2, here we need to learn an $\frac{\epsilon}{nT}$-approx model for each $P_{i,a}$. More so we need to learn an $\frac{\epsilon}{nT}$-approx model of each $P_{i,a}$ with

an error probability of at most $\frac{\delta}{n|A|}$. Also in case of MLeM, each model specifies a distribution over $|A|$ actions. Here, each model $\hat{P}_{i,a}$ specifies a distribution over d items;

Accounting for the above mentioned differences, gives us our new value for m'_k.

$$m'_k = O(\frac{n^5 T^2}{\epsilon^2} log(\frac{n^2 |A| d^{k+1}}{\delta}) \qquad (7.2)$$

This brings us to our main theoretical result concerning LSE-RMAX, stated next.

Lemma 7.2.1. *For any $0 < \epsilon < 1$ and $0 < \delta < 1$, with a high probability of at least $1 - 4\delta$, LSE-RMAX achieves an actual return $\geq U^* - 5\epsilon$ for the general structure learning problem for FMDPs, in a number of time steps given by*

$$O(\binom{n}{2K} \frac{n^6 |A| d^{2K} T^3}{\epsilon^7} log(\frac{n^2 |A| d^{2K+1}}{\delta}) log^2(\frac{1}{\delta})),$$

a quantity polynomial in $\frac{1}{\epsilon}$, $\frac{1}{\delta}$, n^K, d^K, $|A|$ and T.

In comparison the closest competitor of LSE-RMAX from the literature, MET-RMAX requires as input the in-degree K of the transition function's DBN representation. It then generates all possible $\binom{n}{K}$ combinations of DBN representations originating from the possible K choices of parent factors for each factor and tries to learn each of them to a reasonable approximation. After learning them, it uses an elimination mechanism to find the best amongst all of these possible DBN representations for each factor. The sample complexity bound for MET-RMAX for the general structure learning problem is:

$$O(\binom{n}{K} \frac{n^3 |A| K d^K T^3}{\epsilon^6} log(\frac{nK |A| d^{K+1}}{\delta}) log^2(\frac{1}{\delta})),$$

Note the sample complexity bound of MET-RMAX is better than that of LSE-RMAX primarily because it can leverage from prior knowledge of the in-degree K. We conjecture that the sample complexity bound proven for LSE-RMAX is very competitive, if not the best for cases where prior knowledge of K is unavailable. We leave it as an open problem for the readers to improve the bound. However even though LSE-RMAX has a weaker sample complexity bound than MET-RMAX, it seems to work nearly as well as the latter in practice. In the following section, we present empirical results from a couple of benchmark domains demonstrating LSE-RMAX's competitiveness.

7.3 Results

In this section we test LSE-RMAX 's empirical performance in two popular domains compared to its closest competitor MET-RMAX. Theoretically in the context of LSE-RMAX , we need to set the value of m'_k using Equation 7.2. Note that these estimates for m'_k proven theoretically are extremely conservative, and for most practical scenarios, much lower values should suffice. In our experiments we set a fixed m for each k by doing a parameter search over a coarse grid, and choosing the best value. Also we run LSE-RMAX without any exploratory planning iteration. The explorations from the greedy planning iterations were enough to determine the structure of the transition function.

For, MET-RMAX we set K to the in-degree. The m values chosen for the benchmarks are from the results reported in [30]. Additionally for LSE-RMAX we use $\delta = 0.2$, needed for the computation of the σ_k's. All of our results have been averaged over 50 runs.

7.3.1 Stock Trading Domain

Our first domain of interest is the Stock Trading domain introduced in [4]. The domain simulates a stock market composed of a set of economy sectors, each associated with a set of stocks. The size of the domain is defined by the number of sectors (e) and the number of stocks per sector (o). The domain consists of two types of binary variables: e sector ownership variables representing whether or not the agent owns a sector, and $e \times o$ stock variables, representing whether each of the individual stocks is rising or falling. The probability that any given stock will be rising at time $t + 1$ is determined by a combination of the values of all stocks in its sector at time t, according to the formula

$$\Pr(\text{stock rising}) = 0.1 + 0.8 \times \frac{\text{number of stocks in sector rising at time } t}{\text{number of stocks in sector}}$$
$$(7.3)$$

The agent gets a reward of $+1$ for each stock that is rising in a sector that it owns, and -1 for each stock that is not rising. For stocks in sectors that the agent does not own, the reward is 0 regardless of whether they are rising or dropping. The maximum possible reward in a time step is thus $e \times o$, which occurs when the agent owns all sectors and all stocks are rising. The agent's actions are to buy/sell sectors or simply do nothing. Our results are for a domain where $e = 3$ and $o = 2$. For such a setting there are $A = 4$ actions and 36 factors in the factored transition function.

Our results are for LSE-RMAX using $m = 20$. Figure 7.2 shows the cumulative reward plot of both LSE-RMAX and MET-RMAX. Clearly the cumulative reward obtained by both the algorithms are fairly close. In fact the difference is statistically insignificant by a T-test. Figure 7.3 shows the accuracy with which LSE-RMAX performs model selection. The Y-axis gives the fraction of factors for which LSE-RMAX has identified the correct parent factors (correct model) until that time step. Note, that even at time step 1, LSE-RMAX has correctly figured out almost 1/3 of the structure of the transition function.

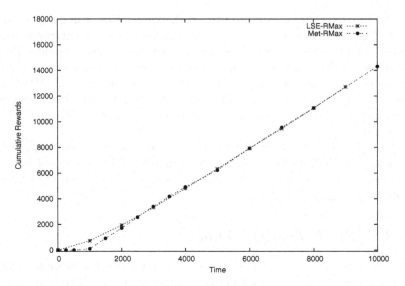

Fig. 7.2 Cumulative reward plot in Stock Trading

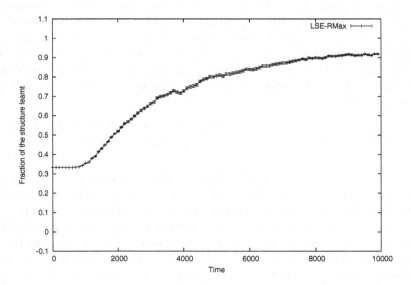

Fig. 7.3 Model selection of LSE-RMAX in Stock Trading

This is because the 3 *e* sector ownership factors have no parent factors and by default LSE-RMAX starts by assuming that each factor is independent of any parent factors. LSE-RMAX figures out the correct structure for almost 60 % of the factors within 2300 time steps, and consistently follows the optimal policy from there onwards. Prior to that LSE-RMAX still accrues a high cumulative reward because it plans based on decent sub-optimal models.

7.3.2 System Administrator Domain

Our second domain of interest is the System Administrator domain introduced in [36]. We use the bidirectional ring topology instance of the problem with 8 machines. The state is represented by 8 binary factors, representing whether or not each of the machines is running. The probability that a machine is running at time $t+1$ depends on whether itself and its two neighbors are running at time t, so the in-degree of the DBN representation for this problem is 3. There are 9 possible actions: reboot the i'th machine, or do nothing. If machine i is down at time t and $reboot(i)$ is executed, it will be running at time $t+1$ with probability 1. If the machine and both its neighbors are running, there is a 0.05 probability that it will fail at the next time step. Each neighbor that is failing at time t increases the probability of failure at time $t+1$ by 0.3. For example, if machine i is running but both neighbors are down, there is a 0.65 chance that machine i will be down at the next time step. Each machine that is running at time t accrues a reward of +1, and if the action taken is reboot there is a penalty of -1.

Here we assume that it is known beforehand that the state of a machine depends on its own state, the action taken, and the states of its K closest neighbors from the past step, K being unknown. So the problem is an instance of the sequential structure learning problem for FMDPs and hence we choose to use LSE-RMAX(S) instead of LSE-RMAX. We also assume that every machine has the same transition structure. Our results are for LSE-RMAX(S) with $m = 30$. Not surprisingly the cumulative reward accrued by LSE-RMAX(S) is significantly better than that of MET-RMAX (see Figure 7.4). The difference

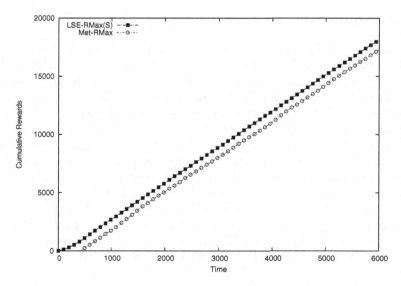

Fig. 7.4 Cumulative reward plot in System Administrator

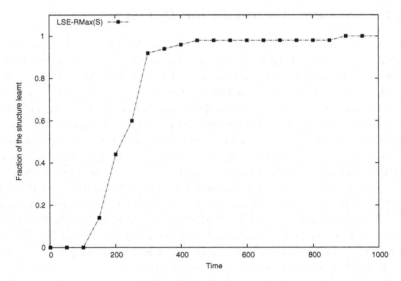

Fig. 7.5 Model selection of LSE-RMAX(S) in System Administrator

is statistically insignificant by a T-test (p-value < 0.05). LSE-RMAX(S) betters MET-RMAX because it makes a more informed structural assumption of the factored transition function than the latter. In this case LSE-RMAX(S) figures out the correct parent factors for all of the factors in about 900 time steps (see Figure 7.5).

7.4   Summary

In this chapter, we showed how our approach to modeling Markovian agents (MLEM, Chapter 5) can be used to solve a challenging problem in RL, namely structure learning in factored MDPs. The current state-of-the-art approaches to the problem require as input the in-degree (K) of the transition function's DBN structure. In this chapter, we were interested in scenarios where prior knowledge of K is unavailable, and K is hard to guess. In this regard we proposed the first structure learning algorithm for ergodic FMDPs, called Learn Structure and Exploit with RMAX (LSE-RMAX), that leverages from our findings in implementing MLEM, to efficiently solve the structure learning problem without requiring a prior knowledge of the in-degree K. Instead we make a different assumption that is analogous to Assumption 1 from Chapters 4 and 5. Since the sample complexity analysis for LSE-RMAX is fairly similar to that of MLEM, our main goal in this chapter was to tie the knot between the sample complexity analysis of the two approaches and derive similar style bounds as in Chapter 5. We argued that our bounds are competitive and compare well with that of MET-RMAX. We also presented empirical results from two popular benchmark domains demonstrating LSE-RMAX's competitiveness.

That concludes all the technical contributions of this book. We now move on to present some related work and in the process situate our research in the literature.

Chapter 8
Related Work

The purpose of this chapter is to situate our research in the literature of Multiagent Learning (MAL). We believe that the algorithms proposed in this book, namely CMLeS and JOMA, extend the frontier of MAL research by a significant margin by achieving a new set of objectives that have not been achieved by any other MAL algorithm to date. Though in this chapter we cite many representative MAL algorithms that are similar in nature to CMLeS and JOMA, we go in depth for only those which are characteristically the closest. In each of these cases wherever applicable, we point out in what ways CMLeS and/or JOMA is different than these algorithms, in the process further underlining the importance of CMLeS and JOMA in the context of MAL research.

While learning in repeated games has been addressed in the field of Game Theory [56, 33] for over a few decades now, the problem received its attention in the Artificial Intelligence (AI) community fairly recently, spanning mostly over the past two decades. The field has since then been called the Multiagent Learning field because of its obvious parallels with the single agent learning field, primarily known as Reinforcement Learning (RL) research [70]. However there is a significant difference between the two problem settings that renders a straightforward extension of RL techniques to solve the MAL problem non-trivial.

Foremost amongst them is understanding what constitutes a good learning goal for the different learning agents in a MAL setting. Unlike the single agent setting, where in most scenarios the underlying setting is a MDP and the objective of the learner is to converge to the optimal policy of the MDP, in a MAL setting the underlying setting is a non-stationary one for which there exists no clear cut optimal policy which the learners can strive to achieve. In such cases, it is very important to define a learning goal for each independent learner which they must strive to achieve as an objective.

Incidentally, it is on this issue that various MAL researchers have disagreed over the years with each having their own preferred criteria that they expect a MAL learner to satisfy. Table 8.1 gives a brief overview of the different scenarios that have been considered in the literature of learning in matrix games. The row and the column indices represent the different agent populations modeled, and some of the different learning criteria that have been proposed in MAL to date, respectively. The entries are representative MAL algorithms that satisfy a particular criterion while interacting with a particular agent population. To the best of our knowledge, the agent populations we have as the row indices are the only ones for which there exist formal learning guarantees. On a similar note, the column indices represent the most popular set of learning criteria proposed.

Table 8.1 situates CMLeS and Joma in the context of MAL research. As is obvious from the table, both CMLeS and Joma satisfy a wide set of

Table 8.1 Multiagent Learning scenarios. The row and the column indices represent the different agent populations modeled, and some of the different learning criteria that have been proposed in MAL to date, respectively. The entries are representative MAL algorithms that satisfy a particular criterion while interacting with a particular agent population. A × entry means that the particular criterion does not apply to the corresponding agent population. A blank entry means there exists no algorithm satisfying that particular criterion for the corresponding agent population.

	Safety	Rationality	Convergence	Maximize SW/ Targeted optimality	Guarded optimality	Universal consistency
self-play	Wolf-Iga, Giga-Wolf, Awesome, Pcm, **CMLeS**, **JOMA**	×	Wolf-Iga, Giga-Wolf, Awesome, **CMLeS**	Pcm, **JOMA**	Pcm, **JOMA**	Giga-Wolf
stationary	Wolf-Iga, Giga-Wolf, Awesome, Pcm, **CMLeS**, **JOMA**	Wolf-Iga, Giga-Wolf, Awesome, Pcm, **CMLeS**, **JOMA**	×	Wolf-Iga, Giga-Wolf, Awesome, Pcm, **CMLeS**, **JOMA**	Wolf-Iga, Giga-Wolf, Awesome, Pcm, **CMLeS**, **JOMA**	Giga-Wolf
memory-bounded	Pcm, **CMLeS**, **JOMA**	×	×	Pcm, **CMLeS**, **JOMA**	Pcm, **CMLeS**, **JOMA**	Giga-Wolf
self, stationary and arbitrary	Pcm, **CMLeS**, **JOMA**,	×	×	×	Pcm	Giga-Wolf
self and mem-bounded	Pcm, **CMLeS**, **JOMA**	×	×	Pcm, **JOMA**	Pcm, **JOMA**	Giga-Wolf
self, mem-bounded, and arbitrary	Pcm, **CMLeS**, **JOMA**	×	×	×	Pcm	Giga-Wolf
self and Markovian	Pcm, **CMLeS**, **JOMA**	×	×	**JOMA**	**JOMA**	Giga-Wolf
self, Markovian, and arbitrary	Pcm, **CMLeS**, **JOMA**	×	×	×		Giga-Wolf
arbitrary	Pcm, **CMLeS**, **JOMA**	×	×	×	Pcm	Giga-Wolf

learning criteria while interacting with different agent populations. CMLeS is the first to achieve convergence, targeted optimality against memory-bounded agents and safety against any other set of agents. CMLeS's guarantee of achieving targeted optimality when every other agent in the population is memory-bounded is also matched by Pcm(A) (denoted as Pcm in Table 8.1), however CMLeS achieves so in a much more sample efficient manner than the latter. On the other hand, Joma is the only MAL algorithm that maximizes SW while interacting with a population comprising Markovian agents. Note, Joma also maximizes SW while dealing with a population comprising memory-bounded agents, and again does so in a much more sample efficient manner than Pcm(A).

We dedicate the rest of this chapter to laying down the different criteria that have been proposed for the MAL learners to satisfy, in the process summarizing key representative algorithms that adhere to each of these criteria. While there is a huge body of literature on MAL where the underlying setting is more complex than a matrix game (such as stochastic games and sequential games [56]), we mostly focus on the literature on learning in matrix games (Sections 8.1 to 8.5): the chosen setting for this book. However in the process we often cite related research pertaining to learning in other complex settings wherever applicable.

8.1 Safety, Consistency and Universal Consistency

Before the MAL problem got attention within the AI community, some of its key challenges were addressed numerous times in Game Theory under the name of Bayes Envelope, dating back to the work of Hannan [37]. In the field of Game Theory, to the best of our knowledge, Fudenberg and Levine [32] were the first to put forth a set of criteria for learning in repeated games. For two player repeated games, their work required the learner to satisfy the following two requirements:- (1) safety: the learning algorithm must guarantee at least the security value of the game; (2) *consistency*: the learning algorithm must guarantee that it does as well as the best response to the empirical distribution of play when interacting against an agent whose policy is stationary. As a follow up they introduced the requirement of *universal consistency* which requires a learning algorithm to perform at least as well as the best response to the historical frequency of play, regardless of the type of policy of the other agent. Note universal consistency means satisfying both safety and consistency, but not vice versa.

They showed that a simple modification of the Fictitious Play algorithm [17] called Cautious Fictitious Play [32], can actually satisfy the criterion of universal consistency. Fictitious Play is a learning algorithm that plays best response based on the historical frequency of actions played by the other agent. Fictitious Play by itself is consistent but not safe. Cautious Fictitious Play builds upon Fictitious Play by choosing each action on a time step with a probability that is an exponential function of that action's expected payoff measured against the historical frequency of the other agent's play. What Fudenberg and Levine show in [32] is that for any policy of the other agent there always exists a value $T > 0$ such that when executed for a period of time $> T$ time steps, Cautious Fictitious Play provably achieves universal consistency.

8.2   Rationality and Convergence

To the best of our knowledge in the AI community, Bowling and Veloso [14] were the first to put forth a set of criteria for evaluating MAL algorithms for repeated matrix games, which was stricter that the ones proposed until then in Game Theory (such as safety, consistency and universal consistency). In matrix games with two players and two actions per player, their proposed algorithm WOLF-IGA [13] satisfies the following criteria:

- Rationality: converges to playing best response against stationary, or *memoryless*, agents. Note rationality is equivalent to the criterion of consistency introduced by Fudenberg and Levine;
- Convergence: converges to playing a Nash equilibrium (NE) joint-policy in self-play;

WOLF-IGA is an improvement over the more general IGA algorithm previously proposed by Singh and Kearns [63], which has very similar guarantees, only failing to converge to NE in one specific scenario of self-play [13].

WOLF-IGA is based on a very simple principle, called the *Win or Learn Fast* principle (or the WOLF principle): "learn quickly while losing, slowly while winning". Each independent WOLF-IGA learner maintains two learning rates, δ_i and δ_j with $\delta_i > \delta_j$. On each time step WOLF-IGA performs gradient ascent on the current policy where the learning rate for performing the gradient ascent depends on the notion of whether the learner is "losing" or "winning". WOLF-IGA considers itself losing when the expected payoff from following the current policy is less than that from following its own share of a pre-chosen NE joint-policy. On the other hand it considers itself winning when the reverse is true. In situations where it is losing, it uses the bigger learning rate δ_i to perform the gradient ascent update. In situations it is winning, it uses the smaller learning rate δ_j. What Bowling and Veloso show in [13] is that if gradient ascent on the current policy is performed in the manner as described above, then WOLF-IGA indeed achieves convergence and rationality.

Subsequent approaches extended the rationality and convergence criteria to arbitrary (multi-player, multi-action) repeated games [7, 27]. Amongst them, AWESOME [27] achieves convergence and rationality in arbitrary repeated games without requiring the agents to observe each others' mixed policy on every time step, while the algorithm by Bannerjee and Peng [7] requires the agents to do so.

AWESOME is very similar in nature to CMLeS. In fact the module of CM-LeS that achieves convergence is inspired from that of AWESOME. Akin to CMLeS, AWESOME assumes access to a NE solver which computes a NE joint-policy for all the agents. The AWESOME learners then attempt to converge to this NE joint-policy by following its own share of the NE joint-policy in repeated phases. Each phase lasts for a duration that monotonically increases with the number of that phase. At the end of each such NE coordination phase, AWESOME checks whether any agent has deviated from following its share of the NE joint-policy by a significant margin. If not, then it starts the next NE coordination phase, otherwise it follows a predefined policy that ensures that if the other agents are indeed stationary, then AWESOME converges to playing best response against them. However again in the process if there is evidence that the other agents are non-stationary, then AWESOME switches back to following the NE coordination phase. Due to this repeated switching between the NE coordination phase and the aforementioned fixed policy, AWESOME provably converges to a NE joint-policy in self-play, or converges to following the best response when the other agents in the population are stationary.

We consider CMLeS to be a significant improvement over both WOLF-IGA and AWESOME as it provides a wider set of guarantees than both. Unlike WOLF-IGA but like AWESOME, CMLeS is designed for arbitrary games. Unlike both WOLF-IGA and AWESOME, CMLeS models a significantly more complex class of agent policies, namely memory-bounded policies, rather than complete stationary polices. Finally CMLeS also ensures that against a set of agents which are neither self nor the target set of memory-bounded agents, it still ensures a return that is close to the security value, i.e., achieves safety. Neither WOLF-IGA nor AWESOME provide any guarantee when the other agent(s) are neither self nor stationary.

8.3 Rationality and No Regret

There has also been significant research on developing MAL algorithms that converge to a NE joint-policy in self-play (achieves convergence) and achieve bounded or no regret against other agents [12, 7, 9] while learning in repeated matrix games. The regret $reg(a_j, s_i)$ of a learning agent i for playing a sequence of actions s_i instead of a playing a fixed action a_j always given that the other agents played the sequence s_{-i}, over T time steps, is defined as follows:

$$reg(a_j, s_i) = \sum_{t=1}^{T} \frac{u_i(a_j, s_{-i}^t) - u_i(s_i^t, s_{-i}^t)}{T} \tag{8.1}$$

s_i^t and s_{-i}^t are the actions played by i and the other agents at time t respectively. Then the *regret* of a learning algorithm over a time period T is defined as $\max_{a_j \in A} reg(a_j, s_i)$. We say a learning algorithm achieves *no regret* when its regret approaches zero as time approaches infinity, when pitted against any agent(s). Note the criterion of achieving no regret is equivalent to the criterion of universal consistency introduced by Fudenberg and Levine.

The most popular amongst algorithms that achieve convergence and no regret is GIGA-WOLF [12] that achieves at most no regret against all other agents and converges to a NE joint-policy in two player two action games. Just like WOLF-IGA builds upon the gradient ascent algorithm IGA, GIGA-WOLF bases itself upon the gradient ascent algorithm GIGA proposed by Martin Zinkevich [77]. GIGA too is inspired from IGA and achieves zero-regret. GIGA-WOLF improves upon GIGA to additionally satisfy the property of convergence.

GIGA-WOLF is a gradient based learning algorithm that internally keeps track of two different gradient updated policies, x_t and z_t . The algorithm chooses actions according to the policy x_t , but updates both x_t and z_t on each time step. The update rules consist of three steps. Step 1 updates x_t according to GIGA's standard gradient update rule and stores the result as x_{t+1}. Step 2 updates z_t in the same manner, but with a smaller step size. Step 3 makes a final adjustment on x_{t+1} by moving it towards z_{t+1}. The magnitude of this adjustment is limited by the change in z_t that occurred in Step 2. We refer the reader to [12] for a detailed account on how these updates are made. What Bowling show in [12] is that if gradient ascent on the current policy is performed in the manner as described above, then GIGA-WOLF indeed achieves convergence and no-regret in two player, two action repeated games.

There is a no clear cut way of comparing GIGA-WOLF with CMLES (or JOMA). Both have their individual theoretical properties which are appealing in their own right . As part of our ongoing research we are trying to incorporate no-reget as a replacement of the safety property for CMLES and JOMA.

8.4   Guarded Optimality with Focus on Targeted Optimality

Most of the algorithms cited in this chapter until now do not generalize well to bigger games and/or more sophisticated agents. For example all of the algorithms from the gradient ascent family (such as WOLF-IGA, IGA, GIGA-WOLF, etc) are mostly tailored for two agent repeated matrix games with two actions per player. AWESOME though generalizes to arbitrary games, but akin to WOLF-IGA and IGA can only deal with stationary agents. Noticing this limitation in the current MAL algorithms for repeated matrix games, more recently Powers and Shoham [59] proposed a new set of evaluation criteria with the hope that algorithms adhering to these new criteria would generalize well to bigger games and against more sophisticated agents. Their criterion called *guarded optimality* requires the learning algorithm to choose a target set of agent behaviors a priori and while interacting in a population comprised of self, agents from this predefined target set, and other arbitrary agents (agents outside the target set) satisfy the following two objectives:

1. achieve close to the SW maximizing joint-return by exploiting the agents from the target set maximally;
2. individually never achieve a return below the security value;

In the special case of two player repeated games, achieving guarded optimality means satisfying the following three objectives:

1. achieve close to SW maximizing joint-return in self-play;
2. when the other agent is drawn from the target set, the learning algorithm should achieve close to the optimal return by exploiting the other agent maximally. This property is popularly known as *targeted optimality*;
3. individually never achieve a return below the security value;

To that end Powers and Shoham propose two algorithms, namely PCM(S) and PCM(A), that achieves guarded optimality against stationary agents (memoryless agents) and the broader class of memory-bounded agents respectively. PCM(A) is an extension of Manipulator [58], a more specific algorithm tailored for 2 player games. Our focus on SW maximization as the target solution concept for the JOMA agents is motivated from the concept of guarded optimality.

Both PCM(S) and PCM(A) are fairly similar in design. We briefly present the algorithmic structure of PCM(S). PCM(A) is similar to PCM(S) except it applies the same learning methodology to learning in the presence of memory-bounded agents. PCM(S) strategizes as follows. To begin with, it follows a "Signal" subroutine for a designated number of time steps. After that it can provably classify all stationary agents into one group and all the PCM(S) agents into another group. The remaining agents are therefore arbitrary agents which are neither PCM(S) nor stationary. If the set of arbitrary agents is empty and there is only one PCM(S) agent in the population, then the corresponding PCM(S) agent switches to playing best response against the identified stationary agents (whose policies have already been learnt from observations in the "Signal" phase). If not, then the PCM(S) agents switch to following a "Coordinate" subroutine that solves a complex optimization problem which ensures that all the PCM(S) agents achieve the SW maximizing joint-return by exploiting the detected stationary agents maximally, while ensuring that they never achieve a return below their security value. Each PCM(S) agent then keeps playing its own share of this joint-policy forever.

Of all the algorithms presented in this book, JOMA is the closest in nature to PCM(S) and PCM(A). JOMA can be seen as an improvement over either of these MAL algorithms in the following ways. First, JOMA deals with a far more complex class of agent behavior (Markovian agents) in comparison to PCM(S) and PCM(A). Second, even while modeling memory-bounded agents, JOMA successfully models memory-bounded agents with a big memory size as long as their policies are based on informative features that transition in a Markovian fashion on every time step, and JOMA's feature set includes those features. In contrast, PCM(A) only models memory-bounded agents of memory size K based on the most recent K joint-actions from history. Since the space of K joint-actions grows exponentially in K, PCM(A) is restricted to modeling memory-bounded agents having a small K. Third, JOMA's sample complexity properties are much more well defined than that of PCM(A) and PCM(S). In case of PCM(A) and PCM(S), the sample complexity bounds are very much dependent on how the internal subroutines (such as "Signal" and 'Coordinate") are implemented. In fact the sample complexity analyses provided for PCM(S) and PCM(A) in [59] are mostly existential in nature. In contrast the sample complexity bounds proven for JOMA are concrete and thorough.

8.5 Focusing Solely on Guarantees in Self-play

Most of the initial research in MAL focuses solely on guarantees in self-play. The assumption is that all agents would follow the proposed MAL algorithm and converge to an initially agreed upon solution concept, such as convergence to a NE joint-policy or maximization of SW. We devote this section to citing some of them.

One of the very popular MAL algorithms in this area is Nash-Q proposed by Hu and Wellman [41]. Nash-Q extends the popular single agent RL algorithm Q-learning [74] to a non-cooperative multiagent context, using the framework of general-sum stochastic games. A *general-sum stochastic game* is a more complex setting than a repeated game. In a general-sum stochastic game, each agent's reward depends on the joint-action of all the agents and the current state, and state transitions obey the Markov property. In other words, each state is a matrix game played by the agents in the system, and transitions between different states following a joint-action is analogous to the transitions in a MDP. A general-sum stochastic game with just one state where all transitions lead back to the same state is a repeated game. Nash-Q maintains Q-functions over joint-actions, and performs updates on them assuming that every agent would exhibit a NE behavior. In order to do so it also maintains Q-functions for other agents (models). The learning protocol provably converges to a NE joint-policy given certain restrictions on the stage games (defined by Q-values) hold. Since a repeated game is a special instance of a general-sum stochastic game, Nash-Q can also be seen as a MAL algorithm for a repeated game.

On a similar line of research focusing on extending Q-learning to a non-cooperative multiagent context, Littman proposed an algorithm Friend-or-Foe Q learning (FFQ) [50] that always converges to a NE joint-policy in self-play. The differences from Nash-Q is that FFQ does not need to maintain Q estimates for the other agents and guarantees convergence to a NE joint-policy in self-play. FFQ can be seen as an extension of the Minimax-Q algorithm [49] also proposed by Littman which focuses solely on two player zero sum games.

Most of the algorithms cited until now in this section use Q-learning as the underlying learning algorithm and use the traditional Value Function based updates on the Q estimates for each agent. On a related note, there is a significant volume of research in MAL that uses Q-learning as the base learning algorithm and does updates based on evolutionary methods, such as the model based on replicator dynamics [44, 72, 7]. A popular representative of such algorithms is the *Frequency Adjusted Q*-learning algorithm, or FAQ-learning [44], which uses a variation of Q-learning that complies with the prediction of an evolutionary model based on replicator dynamics. Though there is no theoretical guarantee of convergence for FAQ-learning, it has been empirically shown to converge to a NE in certain two player two action general sum games.

There has also been some research where the focus has been on convergence to a correlated equilibrium (CE) in self-play rather than a Nash equilibrium [31, 38]. The concept of CE introduced by Aumann [6] can be described as follows: assume on each time step each agent receives a private signal which does not affect the payoffs. The agent then chooses its current step action in the game depending on this signal. A CE of the original game is just a Nash equilibrium of the game with the signals. If the signals are independent across the agents, CE is then just a Nash equilibrium in mixed or pure strategies of the original game. But if the signals are correlated, then it is significantly different than a NE. A good representative of such algorithms is the adaptive procedure proposed by Hart and Mas-Colell [38]. The adaptive procedure is based on the following simple principle. On each time step, continue playing the same action as in the previous time step, or switch to another action with a probability that is proportional to how much higher the cumulative payoff would have been had the agent always made that change in the past. What Hart and Mas-Colell show in [38] is that this simple adaptive procedure generates trajectories of play that almost surely converge to the set of CE (it may not always converge to a unique CE and may oscillate between multiple CE's).

There has also been a significant volume of MAL research that focuses on SW maximization as the target solution concept, rather than convergence to a NE joint-policy. This solution concept though seems more natural for team games or fully cooperative games [56], appears fairly reasonable for general sum games. For example Arieli and Babichenko proposed a learning algorithm for two player games called "Average Testing" [5] that provably achieves close to the social welfare maximizing joint-return by only observing its own payoffs from each step of play. Note, the assumption in this case is much weaker than what we have made in this book. In this book we assume that the matrix game is completely known to all the agents and each agent observes the entire joint-action on each time step. "Average Testing" uses a simple adaptive procedure where each agent sticks to its current step policy if it yields a payoff that exceeds its average payoff by at least some fixed $\epsilon > 0$;

otherwise chooses a policy at random. The trick is when to decide to stick and when to choose randomly.

Works that strive for both SW maximization and convergence to a NE joint-policy are rare. Primarily because it is hard to achieve both in arbitrary games. The most popular amongst them is the algorithm proposed by Littman and Stone that achieves a SW maximizing joint-return in two player repeated games while following a NE action selection strategy [51]. The algorithm is inspired from the ELE algorithm proposed by Brafman and Tennenholtz [16]. The algorithm itself constitutes a policy, and the joint-policy comprising both the agents following the algorithm is a NE joint-policy. The algorithm employs a punishment procedure where any defecting agent (that switches from following the algorithm) is adequately punished (reduced to achieving a payoff at most its security value) so that it becomes irrational for any agent to ever defect. Munoz and Littman later extended the approach to achieve the same in repeated stochastic games [54].

However as suggested earlier, maximizing SW primarily makes sense when the setting is that of a cooperative matrix game, i,e., all agents get the same payoff for each joint-action. Hence most research on cooperative MAL primarily focuses on achieving this solution concept. Some of the early work on cooperative MAL focuses on *joint-action learners*, or JALs [26, 22], which are an extension of Q-learning to cooperative MAL settings. Q-learning in the traditional sense ignores the existence of other agents. Although there has been some research that shows Q-learning can perform well in certain cooperative multiagent settings [62], such results are few and mostly rely on the domain at hand. JALs in contrast, learn the value of each joint-action via integration of RL with equilibrium (or coordination) learning methods. Each Q update for a particular joint-action is similar to that of Q-learning treating a joint-action as a simple action. However for action selection, a JAL learner needs to know what actions the other agents are going to select so that it can coordinate its own action with theirs. For that it maintains beliefs about the models of the other agents, and uses it to coordinate. Subsequent research has extended JAL to arbitrary two player general sum games (such as the work by Bannerjee and Sen [10]) and to more complex team games which are represented in the form of graphical games (such as the works by Guestrin et.al. [35] and Kok and Vlassis [48]).

A problem related to cooperative MAL, is the "Ad-hoc teamwork problem" introduced by Stone et. al. [66]. In this field of research, the goal is to create successful team members, as opposed to successful full teams (because we may not have control over creating the full team). A strong team member must be able to coordinate with its cohorts irrespective of whether the commonality of their algorithms, or behaviors, is substantial or limited. Though the main objective of the 'Ad-hoc teamwork problem" is to look beyond the canonical settings provided by game theory and focus on real world problems, there has been some interesting research that looks at it from the standpoint of learning in a repeated matrix game [67, 2]. In both of these cases, the research has solely focused on how to lead a memory-bounded best response teammate while interacting in a repeated game.

Most of the algorithms cited in this section have their neat theoretical properties catering to their own individual agenda. It is unfair to compare their merits and demerits with the MAL algorithms proposed in this book. Though in this book we decided to look beyond self-play and focus on modeling other adaptive agents, we still relied on certain restrictive assumptions like the underlying framework is that of a matrix game which is completely known beforehand and the actions of all agents on every time step is observable. In this regard, it is important to note that much of the research cited in this section make far weaker assumptions. For example the algorithm by Hart and Mas-Colell that converges to a CE in self-play only relies on observing its own payoff on each time step. Similarly Nash-Q, FFQ and Minimax-Q are tailored for general sum stochastic games, a setting more complex than that of a matrix game.

Finally, though our research mostly studies the MAL problem from a theoretical perspective, there has been some research that studies the MAL problem from an empirical standpoint [3, 11]. The main objective of these works is to empirically study the performance of the different MAL algorithms when pitted against each other and to test their relative strengths and weaknesses in different matrix games. In this regard, this book introduce CMLeS and JoMA as new candidates for study in this way and also provides some initial results to this effect.

There has also been research on opponent modeling where the goal is to propose learning algorithms tailored to exploit specific MAL algorithms. For example Chang and Kaelbling proposed an algorithm [23] that can "fool" Wolf-Phc [14], a variant of Wolf-Iga. Similar observations were noted in [18]. More recently Munoz and Jennings proposed an approach [28] that can exploit Fictitious Play in two player matrix games. Also it is worth noting that there has also been a significant volume of recent work on opponent modeling in different game theoretic settings, apart from repeated matrix games and stochastic games. Most notable amongst them are the works on opponent modeling in Poker, an extensive form game [64, 34]. There has also been some research addressing opponent modeling in games which have emerged from popular competitions, such as the Lemonade Stand game [71] and the games from the Trading Agent competitions [57]. However most of these works are empirical and tailored to the specific domain of interest. As part of our ongoing research, we are seeking avenues for making both CMLeS and JoMA more practically applicable so that they can be applied to solving some of these real world problems.

Having situated CMLeS and JoMA in the literature of MAL algorithms, in the next chapter we present some concluding remarks about our research goals and what we tried to accomplish in this book. In the process, we once again summarize all the contributions of this book and wrap up by citing some short and long term possibilities of future research.

Chapter 9
Conclusion and Future Work

One longstanding goal of AI is the creation of robust autonomous agents that can be deployed in the real world for extended periods of time, without human supervision. Two important capabilities in service of this goal are learning and interaction. Learning is necessary because agent developers cannot be expected to predict the characteristics of all possible environments that the agent might come across in the future. Rather, when situated in a new environment, an autonomous agent must be able to explore the effects of its actions in order to figure out the best possible behavior for its current environment. This challenging problem is well-studied in the area of Reinforcement Learning (RL).

Learning is hard enough when there is only one agent affecting the environment. However, in a world with long-lived fully autonomous agents, these agents may act in the same environment, in which case they will often interact. The area of multiagent systems studies such groups of autonomous interacting entities sharing a common environment. When some or all of these entities are learning, especially about each other, we arrive at the area of Multiagent Learning (or MAL for short).

MAL is often studied in the stylized settings provided by repeated matrix games (normal form games) such as the Prisoner's Dilemma, Chicken and Rock-Paper-Scissors. Repeated games of this type provide the simplest setting that encapsulates many of the key challenges posed by MAL. Specifically, they abstract away the conventional notion of state (situatedness) and allow one to focus purely on the impact of the agents' actions on each other's outcomes, or utilities.

The goal of this book was to develop MAL algorithms that learn to perform better with time against agents which are adaptive. To that end we particularly extend focus to a class of agent behavior that can be modeled as being Markovian. We define a Markovian agent to be one that chooses its actions as a (fixed) function of a set of discrete feature variables computed from the joint history of play, which transition in a Markovian fashion on every time step. As illustrated later in Chapter 2, memory-bounded agents

D. Chakraborty, *Sample Efficient Multiagent Learning in the Presence of Markovian Agents*, Studies in Computational Intelligence 523, DOI: 10.1007/978-3-319-02606-0_9, © Springer International Publishing Switzerland 2014

(agents whose strategy is a function of some historical window of past actions by all the agents [58, 59]) are a special class of Markovian agents whose feature space is the set of joint actions from a bounded history of play.

This book takes a significant step forward in the theory of MAL as it introduces novel algorithms for modeling a comparatively more complex class of agent behavior than has been modeled to date. Furthermore for all of our main algorithms, we provide sample complexity bounds (total number of actions taken to converge to the final desired behavior).

We begin by summarizing the key contributions of this book.

9.1 Summary of the Contributions

In brief the key contributions of this book are as follows:

1. The first contribution concerns modeling memory-bounded agents. In this regard we formally frame the problem as learning in an Adversary Induced Markov Decision Processes (or AIM for short, see Definition 17 from Chapter 2): a general framework for modeling Markovian agents. As part of this contribution, we propose a couple of algorithms that utilize the AIM framework to model memory-bounded agents, assuming their memory size is known. We also present empirical evidence of the effectiveness of these algorithms by pitting them against certain representative algorithms from the MAL and game theory literature.

2. Our second contribution builds on our first contribution, and proposes a novel MAL algorithm called Convergence with Model Learning and Safety (or CMLeS [20] for short) that in a multi-player multi-action (arbitrary) repeated matrix game, is the first to achieve the following three goals:

   - Convergence: converges to following a Nash Equilibrium joint policy profile in self-play (when all other agents are also CMLeS agents);
   - Targeted Optimality against memory-bounded agents: converges to achieving close to the best response with a high certainty, against a set of memory-bounded agents whose memory size is upper bounded by a known value K_{max};
   - Safety: achieves at least the security value against every other set of agents;

   CMLeS is the first of the two major MAL algorithms proposed in this book, the second one is stated as our next contribution.

3. Our third contribution, introduces another novel MAL algorithm called Joint Optimization against Markovian Agents (or JOMA for short), that achieves the following two goals:

   - Maximizes social welfare in the presence of Markovian agents: in the presence of Markovian agents in the population, JOMA provably achieve a joint return very close to the social welfare maximizing joint return (for the JOMA agents), with a high certainty. JOMA assumes prior knowledge of a set of possible features, called the *target set* of features, some of which are assumed to characterize the unknown policy of the Markovian agent. JOMA achieves its objective of modeling the Markovian agents with the most concise model (based on only the relevant features from the target set) via efficient exploration and in the process remains sample efficient.

- Safety: achieves at least the security value against every other set of agents;

Along with a thorough theoretical analysis of JOMA's properties, we also present some empirical results from the gamut test-bed demonstrating its relative effectiveness compared to some of its peers from the MAL literature.

4. Our fourth contribution focuses on a special case scenario of a two player repeated game against a Markovian agent. Here we propose a simplified algorithm, motivated from our findings while implementing JOMA, that efficiently models and exploits a Markovian agent in a two player repeated game. Again we assume prior knowledge of a set of possible features (target set of features) some of which are assumed to characterize the unknown policy of the Markovian agent.

   As part of our empirical analysis for this contribution, we focus on two domains. The first is a challenging new domain called "The Surveillance Game". The game is derived from the multi-robot patrol problem, a well-studied problem in the robotics community, e.g. [1, 53]. The second focuses on "The Ticket Checking problem" introduced in [42]. The problem is inspired by a real life problem of catching passengers who do not buy a ticket while traveling on trains. In both of these domains, to the best of our knowledge, our research takes the first step towards deploying learning agents to solve the problem.

5. The fifth and final contribution of this book shows how our approach of modeling Markovian agents generalizes to solving a broader class of problems pertaining to the single agent RL setting, called structure learning in factored state MDPs (FMDP) [4, 30]. Structure learning is the problem of learning the unknown structures in the underlying transition function of the FMDP from as few samples of online data as possible. Leveraging from our approach of modeling Markovian agents, we propose an alternative mechanism of solving the structure learning problem that results in sample complexity bounds which compare well with those provided by existing approaches. We also show empirically that our approach competes well with the current state-of-the-art structure learning algorithm in certain representative benchmark domains.

That concludes our discussion on all the contributions of this book. Next we highlight some avenues for future research.

9.2   Future Work

We break the segment on future work into two parts. The first part focuses on some near term research possibilities, while the latter on some long term research possibilities.

9.2.1   Near Term Research Possibilities

We present three near term possibilities of future research.

1. The first near term research possibility focuses on looking at avenues that may further tighten our sample complexity bounds. Though we believe that our sample complexity analyses is thorough and the bounds appear to be reasonably tight, it would still be interesting to see whether they can be further tightened. Especially, it would be interesting to check whether a bandit style model selection technique improves our sample complexity bound.

2. An important assumption in this work is that the JOMA agents will always jointly explore to discover the unknown policy of the Markovian agents. Since the current algorithmic framework for JOMA does not punish any agent for not following its prescribed strategy, a self interested agent may be motivated to unilaterally deviate from playing JOMA in search of a higher return for itself. The second near term research possibility focuses on improving JOMA to have a Nash equilibrium exploration policy where no agent has a unilateral incentive to deviate.

3. The third and final near term research possibility involves further empirical testing of our proposed algorithms in suitable domains that are of practical relevance. For example, it might be worthwhile to check how our results on "The Ticket Checking problem" extend to solve the real world problem of catching "evaders". In other words, how TOMMA and TOMMA(S) perform in a real world scenario of catching evaders. In this case, we rely on training and testing TOMMA (and/or TOMMA(S)) on real life data that might be available on request from the Los Angles Metro Rail system authorities.

Next we elaborate on some of the long term research possibilities.

9.2.2   Long Term Research Possibilities

Here we present three long term possibilities of future research.

1. For both CMLeS and JOMA, our default fall back guarantee is safety. That is if there are arbitrary agents (agents which are not memory-bounded in case of CMLeS, or Markovian in case of JOMA) in the population, then each of these algorithms ensure an actual return very close to the security value, with a high certainty. The first long term research possibility focuses on replacing the safety property with the no-regret property (achieve

universal consistency). That is in the presence of arbitrary agents in the population, we require our learning algorithm to achieve no-regret. If such is not achievable, then it would be worthwhile to prove this negative result as an impossibility result.

2. The second long term research possibility focuses on pushing the frontier of agents which can be modeled beyond the Markovian class. It would be particularly worthwhile to discover new classes of agent policies that change over time (non-stationary) and can be modeled. Once we have identified such agent policies, it would interesting to see how JOMA can be extended/modified to include these policies under its umbrella (the umbrella of agent policies that JOMA can successfully model).

3. The final long term research possibility focuses on proposing a practical variant of JOMA that can be applied to solve a real life complex problem. For example creating a Trading Agent Ad Auction (TAC-AA) [43] agent which follows the algorithmic outline of JOMA. We believe this is the most ambitious of all of the research possibilities that we have stated so far. Nonetheless, it is a very interesting and worthwhile direction of future research.

With that we conclude this book. The main goal of this book was to develop MAL algorithms that learn to perform better with time against agents which are adaptive. To that end we particularly extend focus to a class of agent behavior that can be modeled as being Markovian. We believe that the algorithms proposed in this book, namely CMLeS and JOMA (and their key components such as MLeM and others), extend the frontier of MAL research by a significant margin by achieving a new set of goals that have not been achieved by any other MAL algorithm to date. Furthermore for all of our main algorithms, we provide sample complexity bounds wherever applicable. We believe this book takes a significant step forward in the theory of MAL as it introduces novel algorithms for modeling a comparatively more complex class of agent behavior than has been modeled to date.

Appendix A

A.1 Computation of σ_k

For each k, the goal is to select a value for σ_k s.t. Equation. 4.3 in Chapter 4 is satisfied. To repeat, σ_k's are computed such that the following condition is satisfied:

$$\Pr(\Delta_k < \sigma_k) > 1 - \frac{\delta}{K_{max} + 1} \quad \forall k \geq K$$

where δ is a very small constant probability and $\Delta_k \neq -1$.

In the computation of Δ_k, FIND-MODEL chooses a specific $\mathbf{b_k}$, a $\mathbf{b_{k+1}} \in Aug(\mathbf{b_k})$ and an action j for which the models M_k and M_{k+1} differ maximally on that particular time step. Let $M_k(\mathbf{b_k}, j)$ be the probability value assigned to action j by $M_k(\mathbf{b_k})$. Without loss of generality, assume $M_k(\mathbf{b_k}, j) \geq M_{k+1}(\mathbf{b_{k+1}}, j)$. Then $\Delta_k < \sigma_k$ implies satisfying $M_k(\mathbf{b_k}, j) - M_{k+1}(\mathbf{b_{k+1}}, j) < \sigma_k$. For $k \geq K$, we can then rewrite the above inequality as,

$$M_k(\mathbf{b_k}, j) - \mathbb{E}(M_k(\mathbf{b_k}, j)) + (\mathbb{E}(M_{k+1}(\mathbf{b_{k+1}}, j)) - M_{k+1}(\mathbf{b_{k+1}}, j)) < \sigma_k \quad \text{(A.1)}$$

Equation A.1 follows from the reasoning that $\forall k \geq K$, $\mathbb{E}(M_k(\mathbf{b_k}, j)) = \mathbb{E}(M_{k+1}(\mathbf{b_{k+1}}, j))$.

One way to satisfy Inequality A.1 is by ensuring that,

$$M_k(\mathbf{b_k}, j) - \mathbb{E}(M_k(\mathbf{b_k}, j)) < \sigma 1$$
$$\mathbb{E}(M_{k+1}(\mathbf{b_{k+1}}, j)) - M_{k+1}(\mathbf{b_{k+1}}, j) < \sigma 2 \quad \text{(A.2)}$$

and subsequently setting $\sigma_k = \sigma 1 + \sigma 2$.

Now, since we are unsure which pair of $\mathbf{b_k}$ and $\mathbf{b_{k+1}}$, or action may get selected, we need to ensure that the inequalities presented in A.2 are satisfied for all possible choices of $\mathbf{b_k}$, $\mathbf{b_{k+1}}$'s and actions. Thus we need to ensure that the following inequalities are satisfied:

$$\Pr((M_k(\mathbf{b_k}, j) - \mathbb{E}(M_k(\mathbf{b_k}, j)) \geq \sigma 1) \leq \frac{\delta}{2(K_{max} + 1)|A|N_k}, \text{ and}$$

$$\Pr(\mathbb{E}(M_{k+1}(\mathbf{b_{k+1}}, j)) - M_{k+1}(\mathbf{b_{k+1}}, j) \geq \sigma 2) \leq \frac{\delta}{2(K_{max} + 1)|A|N_{k+1}},$$

If the above inequalities are satisfied, then by union bound, we know that that for any pair of $\mathbf{b_k}$ and $\mathbf{b_{k+1}}$, and an action j, both the inequalities presented in A.2 are satisfied with an error probability of at most $\frac{\delta}{K_{max}+1}$. By Hoeffding bound, the above inequalities are always satisfied if we choose

$$\sigma 1 = \sqrt{\frac{1}{2m_k} log(\frac{2(K_{max} + 1)|A|N_k}{\delta})}, \ \sigma 2 = \sqrt{\frac{1}{2m_{k+1}} log(\frac{2(K_{max} + 1)|A|N_{k+1}}{\delta})}$$

Then by subsequently assigning $\sigma_k = \sigma 1 + \sigma 2$, we have our desired result $\Pr(\Delta_k < \sigma_k) > 1 - \frac{\delta}{K_{max}+1}$.

A slightly tighter bound for σ_k can be achieved if we assume that all m_k's are fixed and equal to m (as in our empirical evaluation). If all m_k's are equal to m, then both the estimates $M_k(\mathbf{b_k}, j)$ and $M_{k+1}(\mathbf{b_{k+1}}, j)$ are based on m samples. Then from Hoeffding bound, it follows that:

$$\Pr(|M_k(\mathbf{b_k}, j) - M_{k+1}(\mathbf{b_{k+1}}, j)| < \sigma_k) > 1 - 2exp(-m\sigma_k^2)$$

By bounding $2exp(-m\sigma_k^2)$ by $\frac{\delta}{(K_{max}+1)|A|N_{k+1}}$, we get,

$$\sigma_k = \sqrt{\frac{1}{m} log(\frac{2(K_{max} + 1)|A|N_{k+1}}{\delta})} \tag{A.3}$$

There are $N_{k+1}|A|$ possible ways Δ_k can be computed. Bounding the total error from all such computations using union bound, we get $\Pr(\Delta_k < \sigma_k) > 1 - \frac{\delta}{K_{max}+1}$.

A.2   Proof of Lemma 4.1.1

In this section we present the proof for Lemma 4.1.1 from Chapter 4.

Observation 1: Let at any planning iteration, the probability with which FIND-MODEL selects a model of size $< K$ be p. If all sub-optimal models of size $< K$ are rejected, then it selects $\hat{\pi}_K$ with probability at least $1 - (K_{max} - K)\frac{\delta}{K_{max}+1}$ (from Equation 4.3 main draft and using union bound). Therefore, the probability with which it selects a model $\leq K$ is at least $p + (1 - p)(1 - (K_{max} - K)\frac{\delta}{K_{max}+1}) \geq 1 - \delta$. So models with size $> K$ are only selected with a low probability of at most δ. This is exactly in line with our first goal: we want a model that is at most of size K with a high probability of $1 - \delta$.

Observation 2: If FIND-MODEL selects $\hat{\pi}_K$ as $\hat{\pi}_{best}$, then we have the best model. If it selects a model of size $k < K$, then we have a model which approximates $\hat{\pi}_K$ with an error of at most $\displaystyle\sum_{k \leq k' < K} \Delta_{k'} \leq \sum_{k \leq k' < K} \sigma_{k'}$, over all $\mathbf{b_K}$'s that have been visited m_K times. This follows directly from the definition of Δ_k (Equation 4.2 main draft), and Line 5 of the FIND-MODEL algorithm. The latter ensures that the following is true: $\Delta_k < \sigma_k, \Delta_{k+1} < \sigma_{k+1}, \ldots, \Delta_{K-1} < \sigma_{K-1}$.

Observation 3: Furthermore, from Hoeffding bound, it follows that once a $\mathbf{b_K}$ is visited $O(\frac{1}{\psi^2} log(\frac{N_K|A|}{\delta}))$ times, then $M_K(\mathbf{b_K})$ is a ψ-approx of $\pi_o(\mathbf{b_K})$ with a probability of failure at most $\frac{\delta}{N_K}$. Revisit Equation 4.1 (main draft) for a re-cap on how $\hat{\pi}_K$ is related to M_K.

Observation 4: Combining the above three observations and applying union bound, it follows that once a $\mathbf{b_K}$ is visited $m_K = O(\frac{1}{\psi^2} log(\frac{N_K|A|}{\delta}))$ times, with probability at least $1 - (1 + \frac{1}{N_K})\delta$, $\hat{\pi}_{best}$ is of size at most $k \leq K$ and $\hat{\pi}_{best}(\mathbf{b_K})$ is an $(\displaystyle\sum_{k \leq k' < K} \sigma_{k'} + \psi)$-approx of $\pi_o(\mathbf{b_K})$.

Thus all we need to do is choose m_K such that $\displaystyle\sum_{k \leq k' < K} \sigma_{k'} + \psi$ is bounded by $\frac{\epsilon}{T}$ and ensure that every $\mathbf{b_K}$ gets visited m_K times. It can be shown that the above is satisfied if we choose $m_K = O(\frac{K_{max}^2 T^2}{\epsilon^2} log(\frac{K_{max} N_K |A|}{\delta}))$. The following explains why that is true.

Suppose we choose m_k for any k as follows:

$$m_k = \frac{(2K_{max} + 1)^2 T^2}{\epsilon^2} log(\frac{2(K_{max} + 1)|A|N_k}{\delta})$$

It follows that once a $\mathbf{b_K}$ is visited $m_K = \frac{(2K_{max}+1)^2 T^2}{\epsilon^2} log(\frac{2(K_{max}+1)|A|N_K}{\delta})$ times, then $\hat{\pi}_K(\mathbf{b_K})$ is an $\frac{\epsilon}{(2K_{max}+1)T}$-approx of $\pi_o(\mathbf{b_K})$, with a probability of failure at most $\frac{\delta}{2(K_{max}+1)N_K} < \frac{\delta}{N_K}$ (from Observation 3 by replacing ψ with $\frac{\epsilon}{(2K_{max}+1)T}$ and δ with $\frac{\delta}{2(K_{max}+1)}$).

Assume the worst case that FIND-MODEL returns a model of size 0 ($\hat{\pi}_0$) as $\hat{\pi}_{best}$. Then from Observation 4, this means $\hat{\pi}_{best}(\mathbf{b_K})$ is an $(\sum_{k=0}^{K-1} \sigma_k + \frac{\epsilon}{(2K_{max}+1)T})$-approx of $\pi_o(\mathbf{b_K})$, with probability of failure at most $(1 + \frac{1}{N_K})\delta$.

Now we know,

$$\sigma_k = \sqrt{\frac{1}{2m_{k+1}} log(\frac{2(K_{max} + 1)|A|N_{k+1}}{\delta})} + \sqrt{\frac{1}{2m_k} log(\frac{2(K_{max} + 1)|A|N_k}{\delta})} \quad (A.4)$$

Putting the values of m_k and m_{k+1} in Eqn. A.4 gives us,

$$\forall k, \sigma_k \leq \frac{\sqrt{2}\epsilon}{(2K_{max}+1)T} \tag{A.5}$$

Thus $\sum_{k=0}^{K-1} \sigma_k + \frac{\epsilon}{(2K_{max}+1)T} \leq \frac{\epsilon}{T}$.

So we have shown that once a $\mathbf{b_K}$ is visited m_K times, then $\hat{\pi}_{best}(\mathbf{b_K})$ is an $\frac{\epsilon}{T}$-approx of $\pi_o(\mathbf{b_K})$, with a probability of failure at most $(1 + \frac{1}{N_K})\delta$. The rest of the proof follows from summing up the error from all feasible $\mathbf{b_K}$'s using union bound. Then if follows that once all $\mathbf{b_K}$'s are visited m_K times, the $\hat{\pi}_{best}$ returned by FIND-MODEL is of size at most K and an $\frac{\epsilon}{T}$-approx of π_o with an error probability of at most 2δ.

A.3   Proof of Lemma 4.1.2

In this section we present the proof for Lemma 4.1.2 from Chapter 4. The analysis is an extension of that of RMAX with some differences to account for the learning of an opponent model. We present the proof in steps.

1. The inputs to MLES are K_{max}, δ, ϵ and T. Recall that the (ϵ, T) pair taken as input satisfies Assumption 1. Given an (ϵ, T) pair as input, we need to learn an $\frac{\epsilon}{T}$-approx model of π_o.

   The number of entries denoted as $L1$ that needs to be explored is as follows:

$$L1 = N_K m_K \tag{A.6}$$

   This follows from observing that the size of the relevant state space is N_K and each state needs to be visited m_K times (from Lemma 4.1.1). Now,

$$m_K = O(\frac{K_{max}^2 T^2}{\epsilon^2} log(\frac{K_{max} N_K |A|}{\delta}))$$

   Substituting the value for m_K in Equation A.6, we then get,

$$L1 = O(\frac{N_K K_{max}^2 T^2}{\epsilon^2} log(\frac{K_{max} N_K |A|}{\delta})) \tag{A.7}$$

2. The Implicit Explore and Exploit Lemma of RMAX states that the policy followed by RMAX will either exploit and attain an expected return that is within ϵ of the optimal return for the learned approximate MDP, or will explore with probability at least ϵ in the true MDP (Lemma 6 of [4]). Now we are going to assume the worst case scenario that the explorations to different entries (from Lemma 4.1.1) only happen in the exploratory iterations and when MLES chooses K as the random value from $[0, n]$. For the case of MLES this implies that **eventually at some exploratory iteration**, MLES must choose K as the random value from $[0, K_{max}]$ and also achieve a T-step expected return that is within ϵ of the optimal

return for the approximate MDP. This is because there are finite number of entries to explore (L1 in this case) and hence by the Implicit Explore and Exploit Lemma, RMAX must exploit at some point.

However what we are really concerned about is the return in the true MDP, not in the approximate MDP induced by the learned model. Note, every probability value estimated by our model is $\frac{\epsilon}{T}$ close to the correct value, with a probability of failure at most 2δ (from Lemma 4.1.1). In other words our model is an $\frac{\epsilon}{T}$-approx model of π_o with a failure probability of at most 2δ. Then it follows that the return achieved in the true MDP can never be below 2ϵ of the optimal return U^* over those T steps, with a probability of failure at most 2δ (from the Approximation Lemma of RMax). This follows from the reasoning that if RMAX is exploiting then it must be confining itself to "known" state action pairs (state action pairs for which it believes it has a near accurate model). From Lemma 4.1.1 it is true that the predictions made by $\hat{\pi}_{best}$ for these "known" state action pairs are near accurate with an error of at most $\frac{\epsilon}{T}$.

Now this is where Assumption 1 comes handy. Since, the expected return in the true MDP is at least $U^* - 2\epsilon$, then from Assumption 1, the model $\hat{\pi}_{best}$ based on which we are planning must be sufficient enough to yield the optimal policy. Otherwise such a high expected return would not have been possible. Note no sub-optimal policy could have achieved that high an expected return over T steps. Hence from then onwards in every greedy iteration MLES always follows the optimal policy.

3. For simplification of analysis, assume that the above mentioned exploratory iteration occurs only after all the entries are visited the sufficient number of times and each such entry is only explored in an exploratory iteration where MLES chooses K as the random value from $[0, K_{max}]$. Then the expected number of time steps elapsed before this iteration occurs is,

$$L2 = L1 \times (K_{max} + 1)(\phi + 1) \times \frac{1}{\epsilon} \times T, \text{ where } \phi = \lceil \frac{1-3\epsilon}{\epsilon} \rceil$$

The reasoning behind is as follows:

i. each such iteration in expectation occurs once in every $(K_{max}+1)(\phi+1)$ iterations (since the exploratory iteration happens once every $\phi + 1$ iterations and on each such exploratory iteration a value from $[0, K_{max}]$ is only picked with probability $\frac{1}{K_{max}+1}$;
ii. the exploration probability of visiting a new slot in each such iteration is at least ϵ; and
iii. each iteration lasts for at most T time steps.

Then substituting the values of L1 from Equation A.7 and using $\phi = \lceil \frac{1-3\epsilon}{\epsilon} \rceil = O(\frac{1}{\epsilon})$, we get,

$$L2 = O(\frac{N_K K_{max}^3 T^3}{\epsilon^4} log(\frac{K_{max} N_K |A|}{\delta}))$$

Then from Hoeffding bound, it can be shown that the actual number of time steps taken for all the above explorations to succeed is,

$$L3 = O(L2 \times log(\frac{1}{\delta})) \text{ and } L3 = O(L2 \times log(\frac{1}{\delta})) \qquad (A.8)$$

with a probability of failure at most δ.

4. Once the above has been achieved, from then onwards in every greedy iteration the return is at least $U^* - 2\epsilon$, with a probability of failure at most 3δ. Now since we have an exploratory iteration after every $\phi = \lceil \frac{1-3\epsilon}{\epsilon} \rceil = O(\frac{1}{\epsilon})$ iterations, the expected return over every $\phi + 1$ iterations is at least:

$$\frac{\phi}{\phi+1}(U^* - 2\epsilon) + \frac{1}{\phi+1} \times 0$$
$$\geq U^* - 3\epsilon, \text{ substituting } \phi = \lceil \frac{1-3\epsilon}{\epsilon} \rceil$$

5. Now, in the worst case the expected return over all of the first $L3$ time steps may be 0. This is because the objective of MLES in these time steps is to learn the opponent model to a decent approximation and that often leads to a poor expected return. Thus, the number of time steps needed in total to compensate for the above loss and ensure an expected return of at least $U^* - 4\epsilon$ is at most,

$$L4 = L3 \times \frac{1-3\epsilon}{\epsilon} = O(\frac{L3}{\epsilon}) = O(\frac{L2}{\epsilon}log(\frac{1}{\delta}))$$

Substituting the values of L2 from Equation A.8, we get,

$$L4 = O(\frac{N_K K_{max}^3 T^3}{\epsilon^5} log(\frac{K_{max} N_K |A|}{\delta}) log(\frac{1}{\delta}))$$

6. Finally, what we have shown is that MLES achieves an expected return $\geq U^* - 4\epsilon$, with a high probability of at least $1 - 3\delta$, in L4 time steps. However our aim is to derive a learning time bound for the actual return. Then, by Hoeffding bound, the actual return of MLES is $\geq U^* - 5\epsilon$, with failure probability of at most 4δ, after,

$$L5 = O(\frac{L4}{\epsilon^2}log(\frac{1}{\delta}))$$

number of time steps.

Substituting the values of L4 from Equation A.9, we get,

$$L5 = O(\frac{N_K K_{max}^3 T^3}{\epsilon^7} log(\frac{K_{max} N_K |A|}{\delta}) log^2(\frac{1}{\delta})) \qquad (A.9)$$

This concludes the derivation.

A.4 Achieving Safety When Assumption 1 Does Not Hold

The notion of safety from Fudenberg and Levine [32] requires the learner i to ensure that there always exists a $T > 0$ such that the expected return accrued by i remains $\geq SV_i - \epsilon$ provably for any $T' \geq T$. However for our extended version of MLES that runs in restarts, we show that only at the beginning of any restart, MLES achieves an actual return $\geq SV_i - \epsilon$ with a high certainty. What if the actual return falls below $SV_i - \epsilon$ in every run following a restart? Then we have not achieved safety. In this section we show that provably after a certain number of restarts this never happens.

Now, as a re-cap, each run i lasts for at least the the following time steps:

$$X(i) = O(\frac{N_K K_{max}^3 T_i^3}{\epsilon_i^7} log(\frac{K_{max} N_K |A|}{\delta_i}) log^2(\frac{1}{\delta_i})) \text{ time steps.} \quad (A.10)$$

The values of δ_i, ϵ_i and T_i on run i are assigned as follows:

$$\delta_i = \frac{\delta_{init}}{2^i}, \ \epsilon_i = \frac{\epsilon_{init}}{2^i}, \ T_i = 2^i$$

Substituting these values in Equation A.10, we get,

$$X(i) = O(\frac{N_K K_{max}^3 2^{10i}}{\epsilon_{init}^7} log(\frac{2^i K_{max} N_K |A|}{\delta_{init}}) log^2(\frac{2^i}{\delta_{init}}))$$

$$= O(\frac{N_K^2 K_{max}^4 |A| 2^{13i}}{\epsilon_{init}^7 \delta_{init}^3}) \quad (A.11)$$

In the presence of arbitrary agents in the population (agents who are not K_{max} memory-bounded), MLES converges to modeling them based on some $K \leq K_{max}$. Note, once it switches to a bigger value of K, it cannot go back to a smaller value. Hence in the worst case, MLES converges to modeling them based on a memory size K_{max}. Thus from then onwards each run i lasts for the following time steps,

$$X(i) = O(\frac{N_{K_{max}}^2 K_{max}^4 |A| 2^{13i}}{\epsilon_{init}^7 \delta_{init}^3}) \text{ substituting } K = K_{max} \text{ in Eqn. A.11.}$$

$$= \tilde{O}(2^{13i}) \text{ by getting rid of all the constant terms.} \quad (A.12)$$

Let, $f(x) = 2^{13x}$, and,

$$g(x) = \sum_{j=1}^{x} f(j) - f(x+1) = \sum_{j=1}^{x} 2^{13j} - 2^{13(x+1)} = 2^{\frac{13x(x+1)}{2}} - 2^{13(x+1)} \quad (A.13)$$

Our goal is to show that there exists a value of x such that $g(x)$ is monotonically increasing from that value onwards. Note if that is true, then we know that there exists a value of i, say i', such that from i' and onwards, the difference in the number of time steps elapsed until restart $i' + 1$ and the length of run $i' + 1$ is an increasing function. If that holds, then we must eventually reach a point when we have compensated enough in the preceding runs to ensure that the return never falls below $SV_i - \epsilon$ after the current run, even if the current run yields no return. Hence the rest of the proof focuses on showing that $g(x)$ is an increasing function from some value of x onwards.

Now,

$$\frac{d(g(x))}{dx} = 13log(2)(\frac{2x+1}{2}2^{\frac{13x(x+1)}{2}} - 2^{13(x+1)})$$

which is clearly positive for $x > 2$. Hence by the rule of increasing functions, $g(x)$ is monotonically increasing for $x > 2$. That concludes our analysis.

Appendix B

B.1 Proof of Lemma 5.3.1

In this section we present the proof for Lemma 5.3.1 from chapter 5.

Trivially, if $2K \geq n$, then all n-dimensional feature vector values need to be visited m'_n times. Henceforth, we focus on the more interesting case of $2K < n$.

Since FIND-MODEL-GENERAL incrementally checks for all feature combinations that best describe π_o, at some point it must check whether the correct feature set (F) is the best one. This is when all the sequences checked before it have failed to return a model. It then generates all feature sequences that start with F and there are at most $\sum_{i=1}^{n-K} \binom{n-K}{i} = 2^{n-K} - 1 \leq 2^n$ of them. Observe that if each of the feasible values of the feature set F is visited m_K times, then for every such sequence, the $\hat{\pi}$ returned in Line 10 of FIND-MODEL-GENERAL is the same and an $\frac{\epsilon}{T}$-approx of π_o, with a high probability of at least $1 - 2^n \delta$ (from Lemma 5.2.1 and taking union bound over all such sequences). Thus the $\hat{\pi}_{best}$ returned by FIND-MODEL-GENERAL is an $\frac{\epsilon}{T}$-approx of π_o with a high probability of at least $1 - 2^n \delta$.

However, what if FIND-MODEL-GENERAL returns another model, say $\hat{\pi}_{F'}$, based on feature set F' before it has a chance to test F? Then at some point it must have executed Line 10 on the sequence that starts with the features from F' followed by the ones from F. Observe that the size of this sequence is at most $2K$ (recall $|F| = K$). Assume the worst case that the size is $2K$. Then through a similar reasoning as presented in the previous paragraph, if all the values of the $2K$-dimensional feature vector comprised of features from $F \cup F'$ are visited the number of times specified by Lemma 5.2.1, then $\hat{\pi}_{F'}$ has to be an $\frac{\epsilon}{T}$-approx of π_o, with a high probability of at least $1 - 2^n \delta$. Since we are unsure which F' FIND-MODEL-GENERAL may return, as a safe bet we need to ensure that all the feasible values of any $2K$-dimensional feature vector (originating from combinations of $2K$ features from $\mathcal{F}$) are visited the number of times specified by Lemma 5.2.1, s.t. $\hat{\pi}_{best}$ is an $\frac{\epsilon}{T}$-approx of π_o, with a high probability of at least $1 - 2^n \delta$.

However our goal is to get an $\frac{\epsilon}{T}$-approx of π_o, with a high probability of at least $1 - \delta$. This is achieved by ensuring that all the values of any $2K$-dimensional feature vector is visited m'_{2K} times, where m'_{2K} is much bigger than m_K. We arrive at the value for m'_{2K} by reassigning $\delta \leftarrow \delta/2^n$ in Equation 5.5 and observing that the size of the state space is now the maximum size of the state space from any $2K$-dimensional feature vector (N'_{2K}, not N_K). The substitution from Equation 5.5 that led to the value for m'_{2K} is as follows:

$$O(\frac{n^2 T^2}{\epsilon^2} log(\frac{n N'_{2K} |A| 2^n}{\delta})) \equiv O(\frac{n^3 T^2}{\epsilon^2} log(\frac{n N'_{2K} |A|}{\delta})) = m'_{2K}$$

Then what follows is once all the feasible values of any $2K$-dimensional feature vector are visited m'_{2K} times, the $\hat{\pi}_{best}$ returned by FIND-MODEL-GENERAL is an $\frac{\epsilon}{T}$-approx of π_o, with a high probability of at least $1 - \delta$.

Appendix C

C.1 Computation of σ_k for Equation 6.2

For each k, the goal is to select a value for σ_k s.t. Equation. 6.2 in Chapter 6 is satisfied. To repeat, σ_k's are computed such that the following condition is satisfied:

$$\Pr(\Delta_k \geq \sigma_k) \leq \delta \text{ if } \mathbb{E}(\Delta_k) = 0$$

where δ is a very small constant probability. Our derivation of σ_k in Equation A.3 is very similar to the one presented here with just one difference: there we needed an error probability bound of $\frac{\delta}{n+1}$ instead of δ. Hence the derivation follows in the same line of reasoning as that of the derivation of σ_k in Equation A.3, with the only obvious difference of replacing $\frac{\delta}{n+1}$ by δ. Hence the derived σ_k in this case is,

$$\sigma_k = \sqrt{\frac{1}{m} log(\frac{2|A|N_{k+1}}{\delta})} \tag{C.1}$$

References

1. Agmon, N., Kraus, S., Kaminka, G.A.: Multi-robot adversarial patrolling: Facing a full-knowledge opponent. Journal of AI Research (JAIR) 42, 887–916 (2011)
2. Agmon, N., Stone, P.: Leading ad hoc agents in joint action settings with multiple teammates. In: Proc. of 11th Int. Conf. on Autonomous Agents and Multiagent Systems, AAMAS 2012 (June 2012)
3. Airiau, S., Saha, S., Sen, S.: Evolutionary tournament-based comparison of learning and non-learning algorithms for iterated games. Journal of Artificial Societies and Social Simulation 10, 1–12 (2007)
4. Diuk, C., Strehl, A.L., Littman, M.L.: Efficient structure learning in factored-state mdps. In: AAAI, pp. 645–650 (2007)
5. Arieli, I., Babichenko, Y.: Average testing and the efficient boundary. Discussion Paper Series dp567, Center for Rationality and Interactive Decision Theory, Hebrew University, Jerusalem (February 2011)
6. Aumann, R.: Subjectivity and correlation in randomized strategies. Journal of Mathematical Economics 1(1), 67–96 (1974)
7. Banerjee, B., Peng, J.: Performance bounded reinforcement learning in strategic interactions. In: AAAI 2004: Proceedings of the 19th National Conference on Artifical Intelligence, pp. 2–7. AAAI Press / The MIT Press (2004)
8. Banerjee, B., Peng, J.: Efficient learning of multi-step best response. In: AAMAS 2005: Proceedings of the Fourth International Joint Conference on Autonomous Agents and Multiagent Systems, pp. 60–66. ACM Press, New York (2005)
9. Banerjee, B., Peng, J.: A unifying approach to performance and convergence in online multiagent learning. In: AAMAS 2006: Proceedings of the Fifth International Joint Conference on Autonomous Agents and Multiagent Systems, pp. 798–800. ACM Press, New York (2006)
10. Banerjee, D., Sen, S.: Reaching Pareto Optimality in Prisoner's Dilemma Using Conditional Joint Action Learning. In: Working Notes of the AAAI Workshop on Multiagent Learning (2005)
11. Bouzy, B., Metivier, M.: Multi-agent learning experiments on repeated matrix games. In: Proceedings of the Twenty-Seventh International Conference on Machine Learning (ICML 2010) (June 2010)

12. Bowling, M.: Convergence and no-regret in multiagent learning. In: NIPS 2005: Advances in Neural Information Processing Systems 17, pp. 209–216. MIT Press (2005)
13. Bowling, M., Veloso, M.: Convergence of gradient dynamics with a variable learning rate. In: Proc. 18th International Conf. on Machine Learning, pp. 27–34. Morgan Kaufmann, San Francisco (2001)
14. Bowling, M., Veloso, M.: Rational and convergent learning in stochastic games. In: IJCAI, pp. 1021–1026 (2001)
15. Brafman, R.I., Tennenholtz, M.: R-max - a general polynomial time algorithm for near-optimal reinforcement learning. J. Mach. Learn. Res. 3, 213–231 (2003)
16. Brafman, R.I., Tennenholtz, M.: Efficient learning equilibrium. Artif. Intell. 159(1-2), 27–47 (2004)
17. Brown, G.: Iterative solution to games by fictitious play. In: Activity Analysis of Production and Allocation, pp. 374–376. John Wiley and Sons (1951)
18. Chakraborty, D., Sen, S.: MB-AIM-FSI: A model based framework for exploiting gradient ascent multiagent learners in strategic interactions. In: 7th Int. Conf. on Autonomous Agents and Multiagent Systems (AAMAS), Estoril, Portugal, pp. 371–378 (2008)
19. Chakraborty, D., Stone, P.: Online multiagent learning against memory bounded adversaries. In: Daelemans, W., Goethals, B., Morik, K. (eds.) ECML PKDD 2008, Part I. LNCS (LNAI), vol. 5211, pp. 211–226. Springer, Heidelberg (2008)
20. Chakraborty, D., Stone, P.: Convergence, Targeted Optimality and Safety in Multiagent Learning. In: Proceedings of the Twenty-Seventh International Conference on Machine Learning (ICML 2010) (June 2010)
21. Chakraborty, D., Stone, P.: Structure learning in ergodic factored mdps without knowledge of the transition function's in-degree. In: Proceedings of the Twenty-eighth International Conference on Machine Learning (ICML 2011) (June 2011)
22. Chalkiadakis, G., Boutilier, C.: Coordination in multiagent reinforcement learning: A bayesian approach. In: Proceedings of the Second International Joint Conference on Autonomous Agents and Multiagent Systems, pp. 709–716. ACM Press (2003)
23. Chang, Y., Kaelbling, L.: Playing is believing: the role of beliefs in multi-agent learning. In: NIPS 2001 (2001)
24. Chen, X., Deng, X.: Settling the complexity of two-player nash equilibrium. In: Proceedings of the 47th Foundations of Computer Science (FOCS), pp. 261–272 (2006)
25. Chevaleyre, Y., Dunne, P.E., Endriss, U., Lang, J., Lematre, M., Maudet, N., Padget, J., Phelps, S., Rodrguez-aguilar, J.A., Sousa, P.: Issues in multiagent resource allocation. Informatica 30 (2006)
26. Claus, C., Boutilier, C.: The dynamics of reinforcement learning in cooperative multiagent systems. In: Proceedings of the Fifteenth National/Tenth Conference on Artificial Intelligence/Innovative Applications of Artificial Intelligence, AAAI 1998/IAAI 1998, pp. 746–752. American Association for Artificial Intelligence, Menlo Park (1998)
27. Conitzer, V., Sandholm, T.: Awesome: A general multiagent learning algorithm that converges in self-play and learns a best response against stationary opponents. J. Mach. Learn. Res., 23–43 (2006)

28. de Cote, E.M., Jennings, N.R.: Planning against fictitious players in repeated normal form games. In: Proceedings of the 9th International Conference on Autonomous Agents and Multiagent Systems, AAMAS 2010, vol. 1, pp. 1073–1080. International Foundation for Autonomous Agents and Multiagent Systems, Richland (2010)

29. Degris, T., Sigaud, O., Wuillemin, P.-H.: Learning the structure of factored markov decision processes in reinforcement learning problems. In: ICML 2006: Proceedings of the 23rd International Conference on Machine Learning, pp. 257–264. ACM, New York (2006)

30. Diuk, C., Li, L., Leffler, B.R.: The adaptive k-meteorologists problem and its application to structure learning and feature selection in reinforcement learning. In: ICML 2009: Proceedings of the 26th Annual International Conference on Machine Learning, pp. 249–256. ACM, New York (2009)

31. Foster, D.P., Vohra, R.V.: A randomization rule for selecting forecasts. Oper. Res. 41(4), 704–709 (1993)

32. Fudenberg, D., Levine, D.K.: Consistency and cautious fictitious play. Journal of Economic Dynamics and Control 19(5-7), 1065–1089 (1995)

33. Fudenberg, D., Levine, D.K.: The Theory of Learning in Games (1999)

34. Gilpin, A., Sandholm, T.: A competitive texas hold'em poker player via automated abstraction and real-time equilibrium computation. In: Proceedings of the 21st National Conference on Artificial Intelligence, vol. 2, pp. 1007–1013. AAAI Press (2006)

35. Guestrin, C., Lagoudakis, M.G., Parr, R.: Coordinated reinforcement learning. In: Proceedings of the Nineteenth International Conference on Machine Learning, ICML 2002, pp. 227–234. Morgan Kaufmann Publishers Inc., San Francisco (2002)

36. Guestrin, C., Patrascu, R., Schuurmans, D.: Algorithm-directed exploration for model-based reinforcement learning in factored mdps. In: ICML 2002: Proceedings of the Nineteenth International Conference on Machine Learning, pp. 235–242. Morgan Kaufmann Publishers Inc., San Francisco (2002)

37. Hannan, J.: Approximation to bayes risk in repeated plays. In: Contributions to the Theory of Games, pp. 97–139 (1957)

38. Hart, S., Mas-Colell, A.: A simple adaptive procedure leading to correlated equilibrium. Econometrica 68(5), 1127–1150 (2000)

39. Hester, T., Stone, P.: Generalized model learning for reinforcement learning in factored domains. In: The Eighth International Conference on Autonomous Agents and Multiagent Systems (AAMAS) (May 2009)

40. Hoeffding, W.: Probability inequalities for sums of bounded random variables. Journal of the American Statistical Association, 13–30 (1963)

41. Hu, J., Wellman, M.P.: Nash q-learning for general-sum stochastic games. J. Mach. Learn. Res. 4, 1039–1069 (2003)

42. Jiang, A.X., Yin, Z., Johnson, M., Tambe, M., Kiekintveld, C., Leyton-Brown, K., Sandholm, T.: Towards optimal patrol strategies for fare inspection in transit systems (2012)

43. Jordan, P.R., Wellman, M.P.: Designing an ad auctions game for the trading agent competition. In: IJCAI 2009 Workshop on Trading Agent Design and Analysis, TADA (2009)

44. Kaisers, M., Tuyls, K.: Frequency adjusted multi-agent q-learning. In: Proceedings of the 9th International Conference on Autonomous Agents and Multiagent Systems, AAMAS 2010, vol. 1, pp. 309–316. International Foundation for Autonomous Agents and Multiagent Systems, Richland (2010)
45. Kakade, S.M.: On the sample complexity of reinforcement learning. Phd thesis, University College London (2003)
46. Kearns, M., Koller, D.: Efficient reinforcement learning in factored mdps. In: IJCAI, pp. 740–747 (1999)
47. Kearns, M., Singh, S.: Near-optimal reinforcement learning in polynomial time. In: Proc. 15th International Conf. on Machine Learning, pp. 260–268. Morgan Kaufmann, San Francisco (1998)
48. Kok, J.R., Vlassis, N.: Collaborative multiagent reinforcement learning by payoff propagation. J. Mach. Learn. Res. 7, 1789–1828 (2006)
49. Littman, M.L.: Markov games as a framework for multi-agent reinforcement learning. In: Proceedings of the 11th International Conference on Machine Learning (ML 1994), pp. 157–163. Morgan Kaufmann, New Brunswick (1994)
50. Littman, M.L.: Friend-or-foe q-learning in general-sum games. In: Proceedings of the Eighteenth International Conference on Machine Learning, ICML 2001, pp. 322–328. Morgan Kaufmann Publishers Inc., San Francisco (2001)
51. Littman, M.L., Stone, P.: A polynomial-time nash equilibrium algorithm for repeated games. Decis. Support Syst. 39(1), 55–66 (2005)
52. Mahadevan, S.: Average reward reinforcement learning: Foundations, algorithms, and empirical results. Machine Learning 22 (1996)
53. Marino, A., Parker, L.E., Antonelli, G., Caccavale, F., Chiaverini, S.: A fault-tolerant modular control approach to multi-robot perimeter patrol. In: ICRA (2009)
54. de Cote, E.M., Littman, M.L.: A polynomial-time Nash equilibrium algorithm for repeated stochastic games. In: Proceedings of the Twenty-Fourth Conference on Uncertainty in Artificial Intelligence (UAI 2008), pp. 419–426. AUAI Press, Corvallis (2008)
55. Nash Jr., J.F.: Equilibrium points in n-person games. In: Classics in Game Theory (1950)
56. Osborne, M.J., Rubinstein, A.: A Course in Game Theory. MIT Press, Massachusetts (1994)
57. Pardoe, D., Chakraborty, D., Stone, P.: Tactex09: A champion bidding agent for ad auctions. In: Proceedings of the 9th International Conference on Autonomous Agents and Multiagent Systems (AAMAS 2010) (May 2010)
58. Powers, R., Shoham, Y.: Learning against opponents with bounded memory. In: IJCAI, pp. 817–822 (2005)
59. Powers, R., Shoham, Y., Vu, T.: A general criterion and an algorithmic framework for learning in multi-agent systems. Mach. Learn. 67(1-2), 45–76 (2007)
60. Puterman, M.L.: Markov Decision Processes: Discrete Stochastic Dynamic Programming. Wiley-Interscience (1994)
61. Sela, A., Herreiner, D.K.: Fictitious play in coordination games. Discussion paper serie b, University of Bonn, Germany (1997)
62. Sen, S., Sekaran, M., Hale, J.: Learning to coordinate without sharing information. In: Proceedings of the Twelfth National Conference on Artificial Intelligence, AAAI 1994, vol. 1, pp. 426–431. American Association for Artificial Intelligence, Menlo Park (1994)

63. Singh, S., Kearns, M., Mansour, Y.: Nash convergence of gradient dynamics in general-sum games. In: UAI, pp. 541–548 (2000)
64. Southey, F., Hoehn, B., Holte, R.: Effective short-term opponent exploitation in simplified poker. Machine Learning (2008)
65. Stone, P., Dresner, K., Fidelman, P., Kohl, N., Kuhlmann, G., Sridharan, M., Stronger, D.: The ut austin villa 2005 robocup four-legged team. Technical report, The University of Texas, Austin (2005)
66. Stone, P., Kaminka, G.A., Kraus, S., Rosenschein, J.S.: Ad hoc autonomous agent teams:collaboration without pre-coordination. In: Proceedings of the Twenty-Fourth Conference on Artificial Intelligence (July 2010)
67. Stone, P., Kraus, S.: To teach or not to teach? decision making under uncertainty in ad hoc teams. In: The Ninth International Conference on Autonomous Agents and Multiagent Systems (AAMAS). International Foundation for Autonomous Agents and Multiagent Systems (May 2010)
68. Stone, P., Littman, M.L.: Implicit negotiation in repeated games. In: Meyer, J.-J., Tambe, M. (eds.) Pre-proceedings of the Eighth International Workshop on Agent Theories, Architectures, and Languages, ATAL 2001, pp. 96–105 (2001)
69. Stone, P., Veloso, M.: Multiagent systems: A survey from a machine learning perspective. Autonomous Robots 8(3), 345–383 (2000)
70. Sutton, R.S., Barto, A.G.: Reinforcement Learning. MIT Press (1998)
71. Sykulski, A.M., Chapman, A.C., de Cote, E.M., Jennings, N.R.: Ea2: The winning strategy for the inaugural lemonade stand game tournament. In: ECAI 2010: 19th European Conference on Artificial Intelligence, pp. 209–214. IOS Press, Amsterdam (2010)
72. Tuyls, K., Parsons, S.: What evolutionary game theory tells us about multiagent learning. Artificial Intelligence 171(7), 406–416 (2007)
73. Van Dyke Parunak, H.: Industrial and practical applications of DAI, pp. 377–421 (1999)
74. Watkins, C.J.C.H., Dayan, P.D.: Q-learning. Machine Learning 3, 279–292 (1992)
75. Wolpert, D.H., Macready, W.G.: No free lunch theorems for optimization. IEEE Transactions on Evolutionary Computation 1(1), 67–82 (1997)
76. Wooldridge, M.J.: Introduction to Multiagent Systems. John Wiley & Sons, Inc., New York (2001)
77. Zinkevich, M.: Online convex programming and generalized infinitesimal gradient ascent. In: ICML, pp. 928–936 (2003)

Printed in the United States
By Bookmasters